Transmission Lines in Computer Engineering

Other Computer Engineering Books of Interest

BOSTOCK · *Programmable Logic Devices,* 0-07-006611-6

CHEN · *Computer Engineering Handbook,* 0-07-010924-9

CLEMENTS · *Microprocessor Support Chip Sourcebook,* 0-07-707463-7

CLEMENTS · *68000 Sourcebook,* 0-07-011321-1

DEVADAS, GHOSH, KEUTZER · *Logic Synthesis,* 0-07-016500-9

DEWAR, SMOSNA · *Microprocessors,* 0-07-016639-0

DI GIACOMO · *VLSI Handbook,* 0-07-016903-9

DI GIACOMO · *Digital Bus Handbook,* 0-07-016923-3

ELLIOT · *Integrated Circuit Fabrication Technology,* 0-07-019339-8

HWANG / DEGROOT · *Parallel Processing for Supercomputers and Artificial Intelligence,* 0-07-031606-6

MASSABRIO · *Semiconductor Device Modeling with Spice, Second Edition,* 0-07-02469-3

MUN · *GAAS Integrated Circuits,* 0-07-044025-5

PERRY · *VHDL, Second Edition,* 0-07-049434-7

RAGSDALE · *Parallel Processing,* 0-07-051186-1

SHERMAN · *CD-ROM Handbook,* 0-07-056578-3

SIEGEL · *Interconnection Networks for Large-Scale Parallel Processing,* 0-07-057561-4

SILICONIX, INC · *Designing with Field-Effect Transistors,* 0-07-057537-1

SZE · *VLSI Technology,* 0-07-062735-5

TABAK · *Advanced Microprocessors,* 0-07-062807-6

TRONTELI, TRONTELI, SHENTON · *Analog/Digital ASIC Design,* 0-07-707300-2

TSUI · *LSI/VLSI Testability Design,* 0-07-065341-0

VAN ZANT · *Microchip Fabrication,* 0-07-067194-X

Transmission Lines in Computer Engineering

Sol Rosenstark
Department of Electrical and Computer Engineering
New Jersey Institute of Technology
Newark, New Jersey

McGraw-Hill, Inc.
New York San Francisco Washington, D.C. Auckland Bogotá
Caracas Lisbon London Madrid Mexico City Milan
Montreal New Delhi San Juan Singapore
Sydney Tokyo Toronto

Library of Congress Cataloging-in-Publication Data

Rosenstark, Sol.
Transmission lines in computer engineering / Sol Rosenstark.
p. cm.—(Computer engineering series)
Includes bibliographical references and index.
ISBN 0-07-053953-7
1. Electronic digital computers—Circuits—Design and construction. 2. Microwave wiring. 3. Electric circuit analysis.
I. Title. II. Series.
TK7888.4.R66 1994
621.39'5—dc20 93-41585
CIP

1 2 3 4 5 6 7 8 9 0 DOC/DOC 9 9 8 7 6 5 4 3

ISBN 0-07-053953-7

The sponsoring editor for this book was Steve Chapman and the production supervisor was Suzanne Wardell Babeuf. It was set in Century Schoolbook by Beacon Graphics.

Printed and bound by R. R. Donnelley & Sons Company.

This book is printed on recycled, acid-free paper containing a minimum of 50% recycled deinked fiber.

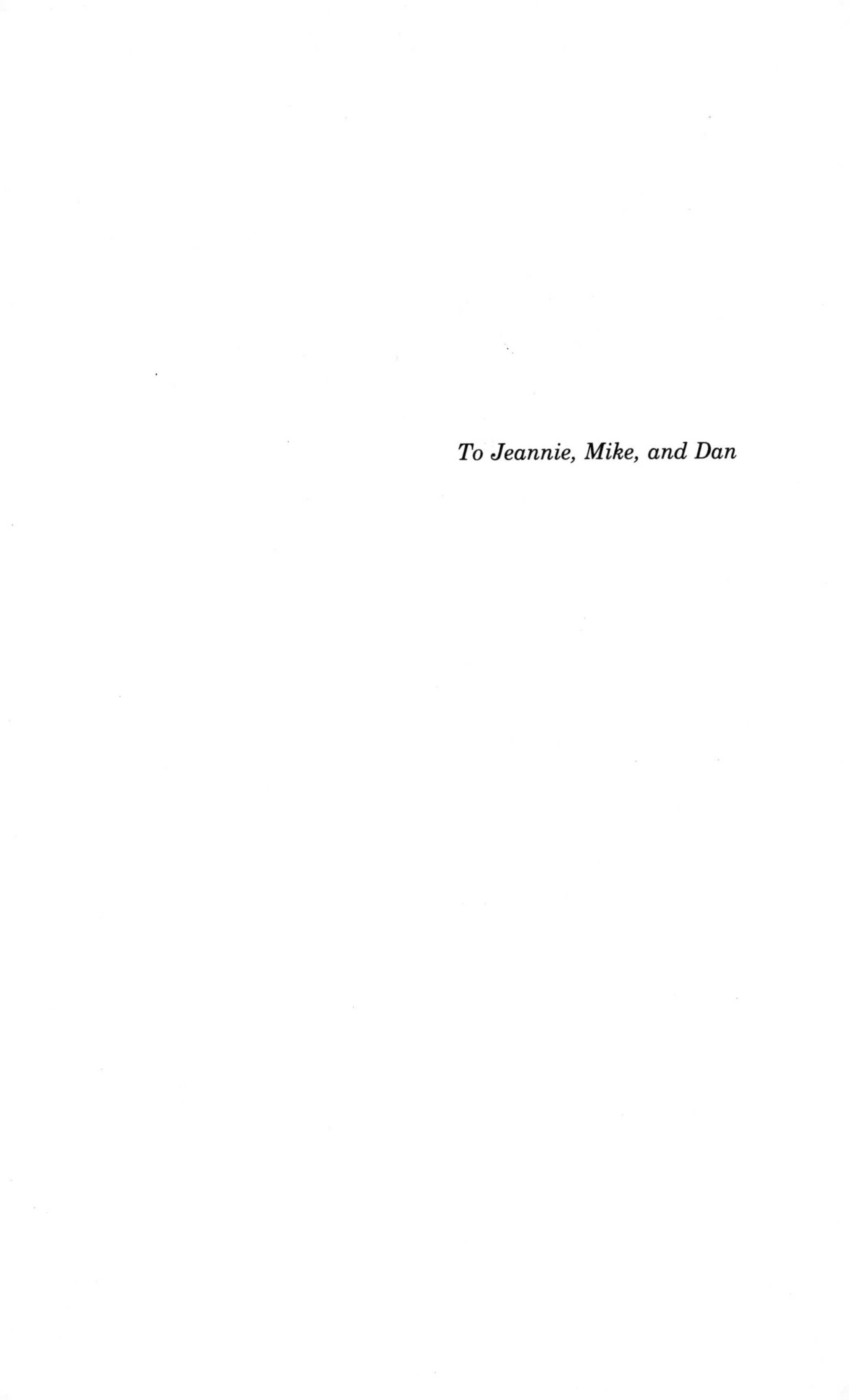

To Jeannie, Mike, and Dan

Contents

Preface

This book is intended for the electrical or computer engineer interested in learning how to interconnect high speed digital circuitry. These interconnection techniques will only grow in importance as available digital circuit devices increase in speed. When the propagation delay through the digital device itself is of the same order of magnitude as the signal transmission-delay between the digital devices, then great care must be exercised in the design of the device interconnections.

Signals propagate through most circuits at or near the speed of light; consequently, a signal transmitted over a transmission line 30 cm (1 foot) in length will experience a delay of 1 ns. This was not a reason for great concern at the time when some of the earlier versions of TTL logic devices had propagation delays of 15 ns. With such slow digital devices, operating on a printed circuit board of 45 cm (1.5 foot) dimension, a 1.5 ns transmission-line delay was of no great consequence. It was correctly assumed that if any reflections did take place, the time required for them to die out would be much less than 15 ns. Hence very little attention was paid to the proper termination of the interconnecting circuitry. However, the same cannot be said of a circuit of the same size designed for the presently available digital devices which have propagation delays of 1 ns or less. Proper precautions must be taken in the design and layout of the device interconnections. This book deals with the analysis and design tools needed to address these problems.

The text itself is equally suitable for either self-study or a formal course. The entire book can be covered in one semester at the graduate level, or selected material can be used for a one-semester upper-level undergraduate course. No special background is required; the only special topic utilized is the Laplace transform. While most upper-level electrical and computer engineering students have previously studied the Laplace transform, a review of this subject is included in Appendix B as a precautionary measure. This book attempts to deal with practical situations and presents only enough mathematical theory when that is indispensable to comprehension of the subject. The text contains many illustrative examples, consequently readers should have no difficulty studying it at their own pace.

This book is organized so that the material follows a clear and logical order of presentation. The first two chapters introduce the subject of transmission-line analysis. Chapter 3 then deals with transients on transmission lines terminated in capacitive loads. Subsequently the very important point is made that it is best to distribute digital loads as uniformly as possible in order to reduce the distortion of pulse shaped signals. Chapter 4 deals with the analysis of non-linear digital drivers and receivers. The so-called Bergeron method is thoroughly explained and applications to FAST TTL and FACT CMOS logic gates are demonstrated.

Chapter 5, which deals with crosstalk on transmission lines, is a thorough explanation of a difficult topic and contains many illustrations. It was my opinion that this topic was in need of a very thorough explanation inasmuch as the presentations found in the literature are very difficult to understand. Many engineers dealing with problems in this area will no doubt welcome a new opportunity to gain a firm grasp of crosstalk fundamentals. Chapter 6 puts the design guidelines for emitter coupled logic (ECL), found in the manufacturers' design handbooks, on a firm footing by supplying the missing derivations.

A book on the subject of transmission lines would not have been complete without a chapter containing material useful for determining transmission-line parameters. Chapter 7 serves that purpose. In all cases, an effort was made to present the most accurate equations to be found in the literature. The temptation existed to present more intricate equations, but those had to be discarded when it was found that the results that they produced were not in agreement with practical reference data.

All chapters contain numerous illustrative examples to facilitate comprehension. Where appropriate, examples are given of circuits which had been analyzed by computer. Where necessary, the required computer programs have been supplied.

I wish to express my gratitude to my colleagues Dr. Edip Niver and Dr. Anthony Robbi for bringing to my attention some material for this book that I would have otherwise missed. My thanks also to Dr. John Carpinelli for teaching me the subtleties of LaTeX. And finally I am indebted to all the students who painstakingly read the manuscript-form course-notes and managed to unearth errors overlooked by me.

Sol Rosenstark

Chapter 1

Transmission Line Fundamentals

1.1 Introduction

Fundamental methods of circuit analysis, such as nodal analysis and mesh analysis, are most useful when applied to lumped parameter circuits. These circuits are constructed of discrete components such as resistors, capacitors, inductors, transformers, transistors, and many other elements too numerous to mention. Such circuits have the distinction that their size is much smaller than the wavelength λ of the signals that are used for their excitation. In a circuit that is purely resistive, a signal applied to one part of the circuit is assumed to be instantly observable at all other parts of the circuit. Delays can be introduced into the circuit by using energy storage devices, such as inductors or capacitors. But it is assumed the wiring itself produces no delays.

Signals propagate through most circuits at or near the speed of light c which in air is 3×10^8 m/s. Consider a fairly large circuit of 30 cm (or 1 foot) in length. We readily find that a signal initiated at one end of this circuit will require 1 ns to propagate to the other end. If the signals that are used in this circuit have a repetition frequency of 60 Hz, with a corresponding period of 16.7 ms, then a 1 ns delay is of no great consequence and it is assumed that an excitation in one part of this circuit manifests itself immediately (with no noticeable delay) in all other circuit parts. But the same cannot be said of a circuit of this size when the signal excitation frequency is in the gigahertz range, since the signal period is of the same order of magnitude as the signal delay. There is another (more direct) way of restating the above argument.

When the exciting signal is of a wavelength that is comparable to the size of the circuit, then there are signal propagation delays within the circuit's wiring which cannot be ignored. To state it purely in terms of wavelength, we use the simple physical relationship between frequency f, wavelength λ, and the speed of light c

$$c = f\lambda \tag{1.1}$$

to determine wavelengths at some interesting frequencies. Using (1.1) we readily find that at audio frequencies ranging between 50 Hz and 20 kHz, the wavelength of signals lies between 6×10^6 m and 15,000 m. These wavelengths are very long in comparison with most circuits, so lumped parameter circuit analysis methods are quite adequate for calculating parameters pertaining to their performance. The same can be said for the video frequency of 10 MHz, at which the wavelength is 30 m (≈100 feet). The wavelength for a 1 GHz wave is 30 cm (≈1 foot), so for circuits whose size is of this order of magnitude, lumped parameter methods are no longer suitable for predicting the behavior of the circuits. We must now take into consideration the fact that signals require time to propagate through the circuits. The methods that we then have to use are referred to as distributed parameter techniques. The reason for this name will soon become apparent.

1.2 Transmission Line Concepts

A section of a two conductor transmission line is shown in figure 1.1. We will first consider a uniform transmission line, meaning that the size of the conductors, their spacing, and dielectric between the two conductors are the same along the entire transmission line.

Signal propagation along transmission lines will be initially considered in terms of voltage and current waves traveling on two uniform

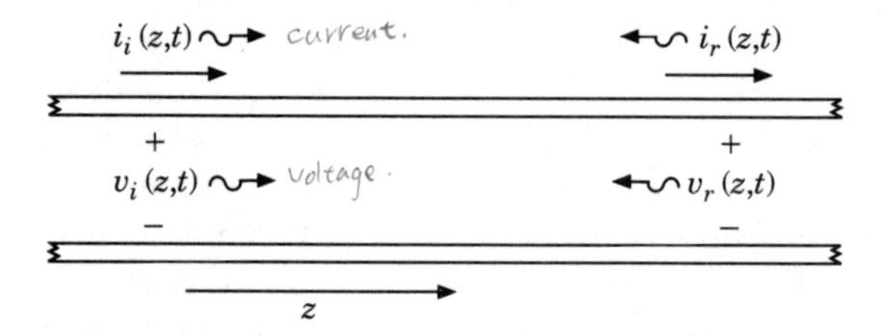

Figure 1.1 A section of a uniform transmission line.

conductors. The voltage and current waves are functions of time t and of the distance along the transmission line z. Figure 1.1 indicates an incident voltage wave $v_i(z,t)$ with positive polarity at the top conductor. The wavy arrow indicates the direction of propagation of this voltage wave. A current wave $i_i(z,t)$ with polarity from left to right is also shown. The wavy arrow in this instance also indicates that the direction of propagation is from left to right. The same convention is used to indicate the reflected voltage wave $v_r(z,t)$ and the reflected current $i_r(z,t)$.

The voltage function $v(z,t)$ and the current function $i(z,t)$ must obey Kirchhoff's laws at all times and at all points along the transmission line. The two conductors of the transmission line are considered to be composed of infinitesimal circuit elements, as shown in figure 1.2.

The transmission line possesses a uniformly distributed series inductance L_0 stated in henries per meter (H/m) and a uniformly distributed series resistance R_0 given in ohms per meter (Ω/m). In addition it possesses a distributed shunt capacitance C_0 given in farads per meter (F/m) and a distributed leakage conductance G_0 stated in siemens per meter (S/m). Figure 1.2 shows the transmission line modeled in terms of infinitesimal repeated sections. As Δz goes to zero, the lumped element line takes on the characteristics of a distributed line.

We use figure 1.3 to derive the equations governing the behavior of the distributed-parameter transmission-line. It shows one section of the repeated structure of figure 1.2 with voltage and current parameters indicated. Since both voltage and current on the two lines depend on two variables, partial derivatives will have to be used with respect to distance z and time t. We arbitrarily assume that all parameters increase as they go from left to right, as shown in figure 1.3. Using Kirchhoff's voltage law, we readily conclude that

$$\Delta v(z,t) = v(z+\Delta z,t) - v(z,t) = -R_0 \Delta z\, i(z,t) - L_0 \Delta z \frac{\partial i(z,t)}{\partial t} \tag{1.2}$$

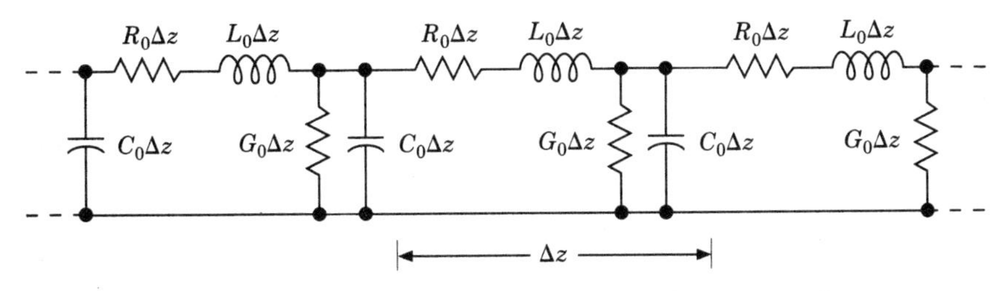

Figure 1.2 A transmission line modeled in terms of lumped parameter elements.

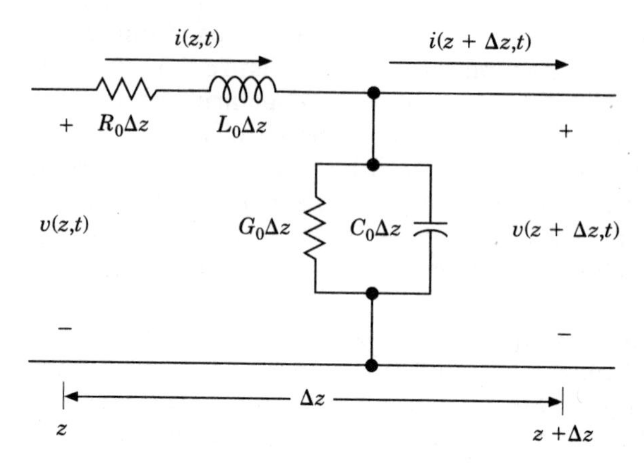

Figure 1.3 One section of the transmission line.

Using Kirchhoff's current law we similarly find that

$$\begin{aligned}\Delta i(z,t) &= i(z + \Delta z, t) - i(z,t) \\ &= -G_0 \Delta z\, v(z + \Delta z, t) - C_0 \Delta z \frac{\partial v(z + \Delta z, t)}{\partial t}\end{aligned} \tag{1.3}$$

The above two expressions are not perfectly symmetrical because the current in the parallel branch is caused by the voltage $v(z + \Delta z, t)$, whereas the voltage drop across the series branch is caused by the current $i(z,t)$. Dividing the above two equations through by Δz and letting Δz approach zero, we obtain the pair of differential equations

$$\lim_{\Delta z \to 0} \frac{\Delta v(z,t)}{\Delta z} = \frac{\partial v(z,t)}{\partial z} = -R_0 i(z,t) - L_0 \frac{\partial i(z,t)}{\partial t} \tag{1.4}$$

$$\lim_{\Delta z \to 0} \frac{\Delta i(z,t)}{\Delta z} = \frac{\partial i(z,t)}{\partial z} = -G_0 v(z,t) - C_0 \frac{\partial v(z,t)}{\partial t} \tag{1.5}$$

The last two expressions are a pair of first order linear partial differential equations with constant coefficients. The equations are linked in that each contains both dependent variables v and i which are functions of the independent variables z and t. It is not possible to solve these equations in general, but solutions are possible for specific cases when particular assumptions are made about the transmission-line parameters R_0, L_0, G_0, and C_0.

The first step toward a solution is to get two equations that are functions of one dependent variable only. To this end we take the partial derivative of (1.4) with respect to z, then change the order of differentiation in the last term to obtain

$$\frac{\partial^2 v(z,t)}{\partial z^2} = -R_0 \frac{\partial i(z,t)}{\partial z} - L_0 \frac{\partial}{\partial t}\left[\frac{\partial i(z,t)}{\partial z}\right] \tag{1.6}$$

We now use (1.5) to replace all references to $\partial i(z,t)/\partial z$ in (1.6) with terms containing only $v(z,t)$, to obtain

$$\frac{\partial^2 v(z,t)}{\partial z^2} = L_0 C_0 \frac{\partial^2 v(z,t)}{\partial t^2} + (L_0 G_0 + R_0 C_0)\frac{\partial v(z,t)}{\partial t} + R_0 G_0 v(z,t) \tag{1.7}$$

A similar procedure is followed to obtain an equation in the variable $i(z,t)$

$$\frac{\partial^2 i(z,t)}{\partial z^2} = L_0 C_0 \frac{\partial^2 i(z,t)}{\partial t^2} + (L_0 G_0 + R_0 C_0)\frac{\partial i(z,t)}{\partial t} + R_0 G_0 i(z,t) \tag{1.8}$$

One should not get the impression that (1.7) and (1.8) are the only equations that the voltage and the current on the transmission line must satisfy. Both variables must still satisfy (1.4) and (1.5) which link the behavior of $v(z,t)$ and $i(z,t)$. In addition, both variables must satisfy boundary conditions, consequently the solutions for both variables will be related. We will now consider some special cases in order to get some insight into possible solutions.

1.3 Traveling Waves on Lossless Lines

It is instructive first to consider transmission lines that have no lossy elements, that is, lines for which $R_0 = G_0 = 0$. There are many instances where lossy lines can be considered lossless. Some examples are lines that are very short in length, so that the effects of losses on the propagation of the signals can be ignored. The operation of many transmission lines can be most easily understood through this idealized theory.

For lossless lines (1.4) and (1.5) reduce to

$$\frac{\partial v(z,t)}{\partial z} = -L_0 \frac{\partial i(z,t)}{\partial t} \tag{1.9}$$

$$\frac{\partial i(z,t)}{\partial z} = -C_0 \frac{\partial v(z,t)}{\partial t} \tag{1.10}$$

In addition (1.7) and (1.8) become

$$\frac{\partial^2 v(z,t)}{\partial z^2} = L_0 C_0 \frac{\partial^2 v(z,t)}{\partial t^2} \tag{1.11}$$

$$\frac{\partial^2 i(z,t)}{\partial z^2} = L_0 C_0 \frac{\partial^2 i(z,t)}{\partial t^2} \tag{1.12}$$

It can be verified by substitution that any function of the type

$$g_+(z,t) = g\left(t - \sqrt{L_0 C_0}\, z\right) \tag{1.13}$$

or of the type

$$f_-(z,t) = f\left(t + \sqrt{L_0 C_0}\, z\right) \tag{1.14}$$

will satisfy both (1.11) and (1.12). We want to see if the expressions given above have any specific interpretation.

At some arbitrary time t_0, (1.13) evaluates for all z to

$$g_+(z,t_0) = g\left(t_0 - \sqrt{L_0 C_0}\, z\right) \tag{1.15}$$

A short time later at time $t_0 + \delta t$ (1.13) evaluates to

$$g_+(z,t_0 + \delta t) = g\left(t_0 + \delta t - \sqrt{L_0 C_0}\, z\right) \tag{1.16}$$

which can be rewritten into the form

$$g_+(z,t_0 + \delta t) = g\left[t_0 - \sqrt{L_0 C_0}\left(z - \frac{\delta t}{\sqrt{L_0 C_0}}\right)\right] \tag{1.17}$$

For simplicity, a simple waveform was arbitrarily chosen for $g_+(z,t)$, then (1.15) and (1.17) were plotted in figure 1.4. It is apparent from (1.17) and the illustration in figure 1.4 that the waveform on the transmission line at time $t_0 + \delta t$ is identical to that at time t_0 except that it appears displaced to the right by the distance

$$\delta z_+ = \frac{\delta t}{\sqrt{L_0 C_0}} \tag{1.18}$$

We therefore conclude that (1.13) is the equation for a wave traveling on the transmission line in the direction of increasing z. The speed

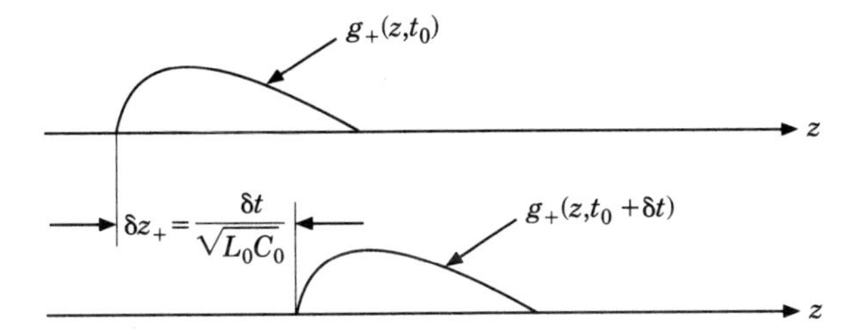

Figure 1.4 An arbitrary waveform $g_+(z,t)$ shown at two time instants.

of propagation ν_+ of this wave, which travels in the positive going z direction, can be deduced from (1.18) by rearranging and taking the limit as δt goes to 0.

$$\nu_+ = \lim_{\delta t \to 0} \frac{\delta z_+}{\delta t} = \frac{1}{\sqrt{L_0 C_0}} \tag{1.19}$$

Applying a procedure identical to that carried out in steps (1.15) to (1.19) to the function $f_-(z,t)$ as given in (1.14), we arrive at the wave propagation diagram in figure 1.5. The waveform on the transmission line at time $t_0 + \delta t$ is identical to that at time t_0 except that it is displaced to the left by a distance $\delta t/\sqrt{L_0C_0}$. This is the same as saying that it is displaced to the right by a negative distance given by

$$\delta z_- = -\frac{\delta t}{\sqrt{L_0 C_0}} \tag{1.20}$$

We therefore conclude that (1.14) is the equation for a wave traveling on the transmission line in the direction of decreasing z. The speed of propagation ν_- of this wave, which travels in the negative going z

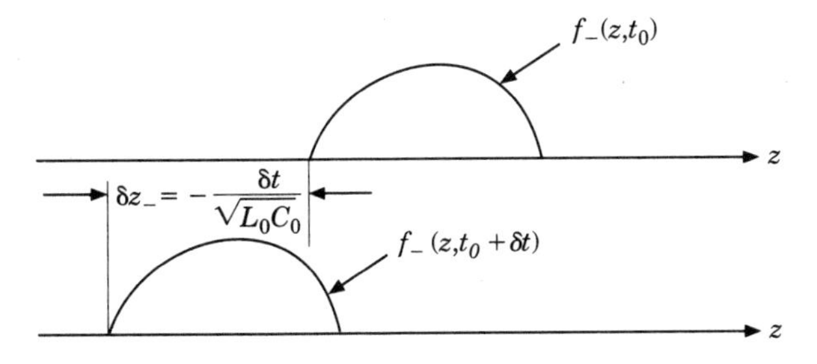

Figure 1.5 An arbitrary waveform $f_-(z,t)$ shown at two time instants.

direction, can be deduced from (1.20) by rearranging and taking the limit as δt goes to 0.

$$\nu_- = \lim_{\delta t \to 0} \frac{\delta z_-}{\delta t} = -\frac{1}{\sqrt{L_0 C_0}} \tag{1.21}$$

We will henceforth unify the speed concepts of (1.19) and (1.21) by using one positive quantity

$$\nu \equiv |\nu_+| = |\nu_-| = \frac{1}{\sqrt{L_0 C_0}} \tag{1.22}$$

At this point it is timely to consider the units of the speed ν. Inductance and capacitance are defined in terms of the physical parameters that they relate. Since inductance relates voltage to the rate of change of current in an inductance, then the henry is in fact a volt-sec/amp. Similarly capacitance relates current to the rate of change of voltage, so that the farad is in fact an amp-sec/volt. All the transmission-line parameters introduced at the beginning of this chapter are per-unit-length parameters, so that L_0 is given in units of H/m and C_0 is given units of F/m. We can therefore use the equalities

$$\text{units of } L_0 = \frac{\text{H}}{\text{m}} = \frac{\text{volt-sec}}{\text{amp-m}} \tag{1.23}$$

$$\text{units of } C_0 = \frac{\text{F}}{\text{m}} = \frac{\text{amp-sec}}{\text{volt-m}} \tag{1.24}$$

to find the units of calculated transmission-line parameters.

Example 1.1 The distributed capacitance and inductance parameters of some cable are specified as

$$C_0 = 70\,\text{pF/m} \tag{1.25}$$

and

$$L_0 = 0.35\,\mu\text{H/m} \tag{1.26}$$

The product $L_0 \times C_0$ is now calculated with the aid of the unit conversions given in (1.23) and (1.24) with the result

$$L_0 C_0 = 2.45 \times 10^{-17} \qquad (\text{s/m})^2$$

Using this in (1.22) we find that the signal propagates on the cable at the speed

$$\nu = 2.02 \times 10^8 \qquad (\text{m/s}) \tag{1.27}$$

which is approximately 67% of the speed of light in a vacuum and is in the range of propagation speeds for some commercial cables. ■

1.4 The Characteristic-Impedance Concept

Let us suppose that we have on a transmission line a voltage waveform $v(z,t)$ composed of a positive and negative going wave

$$v(z,t) = v_+(z,t) + v_-(z,t) \tag{1.28}$$

as shown in figure 1.6. The positive and negative going voltages appearing on the right side of the above equation are respectively given by

$$v_+(z,t) = g\left(t - \sqrt{L_0C_0}\,z\right) \tag{1.29}$$

$$v_-(z,t) = f\left(t + \sqrt{L_0C_0}\,z\right) \tag{1.30}$$

The above voltages are assumed to have positive polarity, as shown in figure 1.6.

We will similarly assume the current on the line to be the sum of a positive and negative going current, as shown in figure 1.6.

$$i(z,t) = i_+(z,t) + i_-(z,t) \tag{1.31}$$

Substituting (1.29) and (1.30) into (1.28), then substituting that into (1.9), we obtain without too much difficulty the result

$$i(z,t) = \sqrt{\frac{C_0}{L_0}}\,g\left(t - \sqrt{L_0C_0}\,z\right) - \sqrt{\frac{C_0}{L_0}}\,f\left(t + \sqrt{L_0C_0}\,z\right) \tag{1.32}$$

The quantity $\sqrt{L_0/C_0}$ is commonly referred to as the characteristic impedance of the lossless line designated by the symbol Z_0, so that

$$Z_0 \equiv \sqrt{\frac{L_0}{C_0}} = \text{characteristic impedance} \tag{1.33}$$

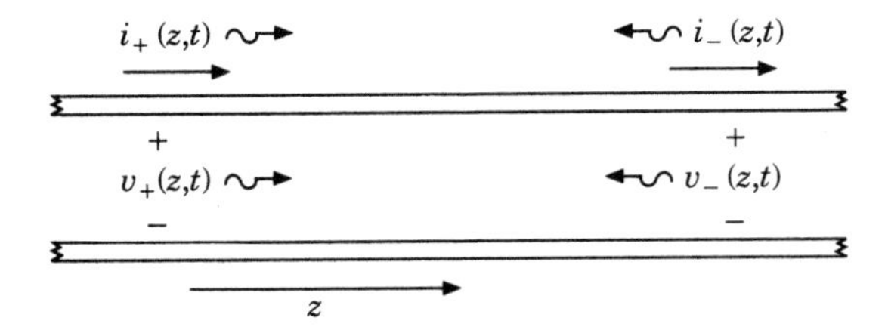

Figure 1.6 A lossless line with bidirectional voltage and current waveforms.

and from the discussion pertaining to (1.23) and (1.24) it is apparent that the characteristic impedance must have dimensions of ohms.

Using (1.33) and comparing (1.32) with (1.31), we find that

$$i_+(z,t) = \frac{1}{Z_0} g\left(t - \sqrt{L_0 C_0}\, z\right) \tag{1.34}$$

$$i_-(z,t) = -\frac{1}{Z_0} f\left(t + \sqrt{L_0 C_0}\, z\right) \tag{1.35}$$

Comparing (1.34) and (1.35) with (1.29) and (1.30) we see that

$$i_+(z,t) = \frac{1}{Z_0} v_+(z,t) \tag{1.36}$$

$$i_-(z,t) = -\frac{1}{Z_0} v_-(z,t) \tag{1.37}$$

It is now clear that the positively traveling voltage wave $v_+(z,t)$ is accompanied by a current wave $i_+(z,t)$. They both propagate from left to right, as shown in figure 1.6. If there is a negatively traveling voltage wave $v_-(z,t)$, then this is accompanied by a current wave $i_-(z,t)$. They both propagate from right to left, as shown in figure 1.6.

Example 1.2 The distributed capacitance and inductance parameters of some cable are the same as those specified in example 1.1. We find that the characteristic impedance of the line calculated using 70 pF/m for C_0 and 0.35 μH/m for L_0 is

$$Z_0 = \sqrt{\frac{L_0}{C_0}} = \sqrt{\frac{0.35 \times 10^{-6}}{70 \times 10^{-12}}} = 70.7\,\Omega \tag{1.38}$$

Commercial cables possessing a characteristic impedance of 50 Ω, 75 Ω, 93 Ω, and 125 Ω are very commonplace. ■

Example 1.3 A coaxial transmission line consisting of an inner conductor of diameter d, separated by a dielectric from an outer conductor whose inside diameter is D, is shown in figure 1.7. It can be readily found (in most electromagnetic-theory texts) that the distributed capacitance of such a cable is given by

$$C_0 = \frac{2\pi\epsilon}{\ln(D/d)} \text{ (F/m)} \tag{1.39}$$

and the distributed inductance is

$$L_0 = \frac{\mu}{2\pi} \ln(D/d) \text{ (H/m)} \tag{1.40}$$

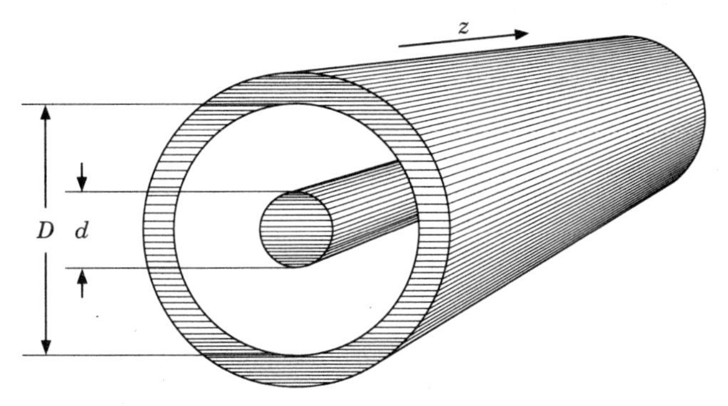

Figure 1.7 Geometry of a coaxial cable.

In the above expressions ϵ is the dielectric permittivity and μ is the magnetic permeability of the dielectric. For a vacuum they are

$$\epsilon_0 \approx \frac{1}{36\pi} \times 10^{-9} \text{ (F/m)} \tag{1.41}$$

$$\mu_0 = 4\pi \times 10^{-7} \text{ (H/m)} \tag{1.42}$$

For other materials they can be calculated by using the relative permittivity ϵ_r to find $\epsilon = \epsilon_r \epsilon_0$ and the relative permeability μ_r to find $\mu = \mu_r \mu_0$. Since dielectrics are made of non-magnetic materials (usually plastics and sometimes air or an inert gas), the dielectrics possess the relative permeability of free space. So if a specification for μ_r is not given, it is usually taken as 1.

Using (1.39) and (1.40) we readily find that for a coaxial transmission-line

$$\nu = \frac{1}{\sqrt{\mu\epsilon}} \tag{1.43}$$

and

$$Z_0 = \frac{1}{2\pi}\sqrt{\frac{\mu}{\epsilon}} \ln\left(\frac{D}{d}\right) \tag{1.44}$$

For the values of μ and ϵ given in (1.41) and (1.42) it is readily found that the speed of wave propagation ν is 3×10^8 m/s, which is the speed of light in free space. ■

We are now in a position to summarize the important results for lossless transmission-lines.

Summary of Important Results for Lossless Lines The lossless transmission-line has voltage and current waves with the directions of travel and polarities shown in figure 1.6. The voltage waveform $v(z, t)$ is composed of positive going (incident) and negative going (reflected) waves.

$$v(z,t) = v_+(z,t) + v_-(z,t) \tag{1.45}$$

With the voltage waveform is associated a current waveform composed of incident and reflected current waves.

$$i(z,t) = i_+(z,t) + i_-(z,t) \tag{1.46}$$

The two voltage waves can be expressed in the general form

$$v_+(z,t) = g(t - z/\nu) \tag{1.47}$$

and

$$v_-(z,t) = f(t + z/\nu) \tag{1.48}$$

The above voltage waves are always accompanied by current waves. The voltages and currents are related by the characteristic impedance Z_0 as given below

$$i_+(z,t) = \frac{1}{Z_0} v_+(z,t) \tag{1.49}$$

$$i_-(z,t) = -\frac{1}{Z_0} v_-(z,t) \tag{1.50}$$

The quantity Z_0 which relates the voltage and current waves is the characteristic impedance of the lossless line and is given by

$$Z_0 = \sqrt{\frac{L_0}{C_0}} \tag{1.51}$$

For lossless transmission-lines the speed of propagation that appears in (1.47) and (1.48) is given by

$$\nu = \frac{1}{\sqrt{L_0 C_0}} \tag{1.52}$$

■

1.5 Sinusoidal Waves — Power Propagation

It is known for linear lumped parameter circuits that sinusoidal excitations produce sinusoidal responses of the same frequency. Transmission lines are a generalization of lumped linear circuits, so the same is true in this situation as well. We will find it instructive to examine the behavior of lossless lines in this situation to see if we can gain some insight into the power carrying properties of the waves.

For sinusoidal time dependence the general function (1.47) can be replaced with the more specific

$$v_+(z,t) = V_i \cos[\omega(t - t_1 - z/\nu)] \tag{1.53}$$

In the above equation V_i is the peak amplitude of the voltage, and the time t_1 represents an arbitrarily chosen time delay to reflect the fact that the starting time of the wave does not in any way affect the result. We can similarly replace the general function (1.48) by

$$v_-(z,t) = V_r \cos[\omega(t - t_2 + z/\nu)] \tag{1.54}$$

When the last two equations are substituted into (1.45), we have the total voltage on the line

$$v(z,t) = V_i \cos[\omega(t - t_1 - z/\nu)] + V_r \cos[\omega(t - t_2 + z/\nu)] \tag{1.55}$$

Using the fact that voltages and currents are related by Z_0 as stated in (1.49) and (1.50), we get

$$i_+(z,t) = \frac{V_i}{Z_0} \cos[\omega(t - t_1 - z/\nu)] \tag{1.56}$$

and

$$i_-(z,t) = -\frac{V_r}{Z_0} \cos[\omega(t - t_2 + z/\nu)] \tag{1.57}$$

When the last two equations are substituted into (1.46), we have the total current on the line

$$i(z,t) = \frac{V_i}{Z_0} \cos[\omega(t - t_1 - z/\nu)] - \frac{V_r}{Z_0} \cos[\omega(t - t_2 + z/\nu)] \tag{1.58}$$

It is instructive to find an expression for the power on the line which the voltage wave of (1.55) produces when interacting with the current wave of (1.58). We simply multiply the two expressions to find that

$p(z,t)$, the instantaneous power distribution on the line, is composed of two terms

$$p(z,t) = \frac{V_i^2}{Z_0}\cos^2[\omega(t - t_1 - z/\nu)] - \frac{V_r^2}{Z_0}\cos^2[\omega(t - t_2 + z/\nu)] \tag{1.59}$$

We observe that the power on the line is the difference of an incident and a reflected power term, namely

$$p(z,t) = p_+(z,t) - p_-(z,t) \tag{1.60}$$

and we can identify each by comparing (1.60) with (1.59) as

$$p_+(z,t) = \frac{V_i^2}{Z_0}\cos^2[\omega(t - t_1 - z/\nu)] \tag{1.61}$$

and

$$p_-(z,t) = \frac{V_r^2}{Z_0}\cos^2[\omega(t - t_2 + z/\nu)] \tag{1.62}$$

It is apparent from (1.61) that we have a power wave traveling on the line in the positive z direction and that the instantaneous power on the line is a function of both distance z and time t. At any instant of time t, it is sinusoidally dependent on the distance z. We can also say that at any point z on the transmission line, the power is sinusoidally dependent on the time t. The sinusoidal dependence is also true for (1.62) except that this power wave is traveling in the negative z direction.

We can readily find the time average of the instantaneous power expression of (1.61) with the result

$$P_+(z) = \frac{V_i^2}{2Z_0} \tag{1.63}$$

The above is the time averaged power carried by the incident wave. Repeating the averaging process for the reflected power term of (1.62) we obtain

$$P_-(z) = \frac{V_r^2}{2Z_0} \tag{1.64}$$

The above is the time averaged power carried by the reflected wave. From (1.60) we conclude that the total averaged power must be given by

$$P(z) = P_+(z) - P_-(z) \qquad (1.65)$$

The choice of a sinusoidal wave for the analysis was made because it is an easy and simple waveform which makes calculation easy. It gave us some insight into the fact that we are not simply dealing with the transmission of voltage and current but that those waves in fact are responsible for the transmission of power.

1.6 Distortionless Transmission

In figure 1.8 we have an input signal $x(t)$ applied to a transmission medium. The question is, "What kind of characteristics should the transmission medium possess so that it will be considered distortionless?" It is generally accepted that a distortionless medium should produce a signal $y(t)$ at the output which is a delayed and scaled replica of the input signal $x(t)$. In other words

$$y(t) = ax(t - t_d) \qquad (1.66)$$

The time duration t_d is the time delay experienced by the input signal as it passes through the transmission medium. The constant a represents a scale factor. For an active device, such as an amplifier, it can be greater than unity. For a passive medium, such as a transmission line, a is less than or equal to 1. So it should be clear that a device does not have to be lossless to be distortionless.

When the Laplace transform is applied to (1.66), we find that

$$Y(s) = aX(s)e^{-t_d s} \qquad (1.67)$$

Hence the transfer function $H(s)$ for this medium, which is the ratio of the Laplace transform of the output signal $Y(s)$ to the Laplace transform of the input signal $X(s)$, is

$$H(s) = \frac{Y(s)}{X(s)} = ae^{-t_d s} \qquad (1.68)$$

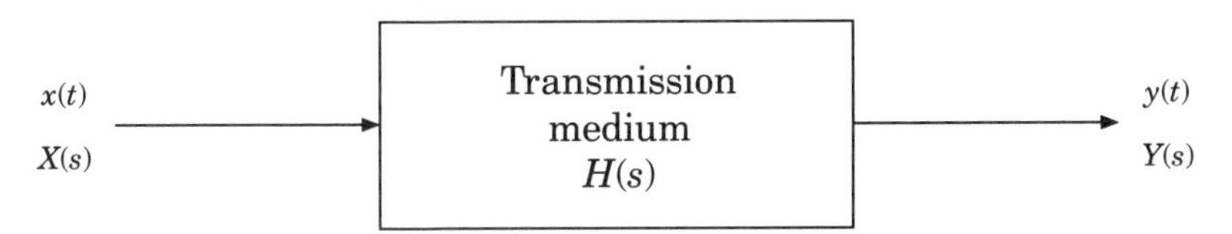

Figure 1.8 Transmission medium with input and output.

Whereas (1.66) states the requirements for distortionless transmission in the time domain, (1.68) does the same thing in the s-domain.

1.7 Distortionless Lossy Lines

A transmission line whose parameters satisfy

$$\frac{R_0}{L_0} = \frac{G_0}{C_0} \tag{1.69}$$

is known as a distortionless (but not lossless) transmission line. When the above holds, we can show by substitution that solutions of the type

$$g_+(z,t) = e^{-R_0\sqrt{C_0/L_0}\,z}\,g\left(t - \sqrt{L_0C_0}\,z\right) \tag{1.70}$$

and

$$f_-(z,t) = e^{+R_0\sqrt{C_0/L_0}\,z}\,f\left(t + \sqrt{L_0C_0}\,z\right) \tag{1.71}$$

satisfy (1.7) and (1.8).

The terms $g(\cdot)$ in (1.70) and $f(\cdot)$ in (1.71) are expressions that account for pure propagation. In addition, each waveform of the last two equations has an attenuation factor embodied in the exponential term. As $g_+(z,t)$ propagates to the right it decreases in size, but the waveshape remains undistorted. Similarly, as $f_-(z,t)$ propagates to the left it decreases in size, but its waveshape also remains undistorted. This is illustrated in figure 1.9.

In (1.70) and (1.71) we define the term

$$\alpha \equiv R_0\sqrt{\frac{C_0}{L_0}} = \text{attenuation constant} \tag{1.72}$$

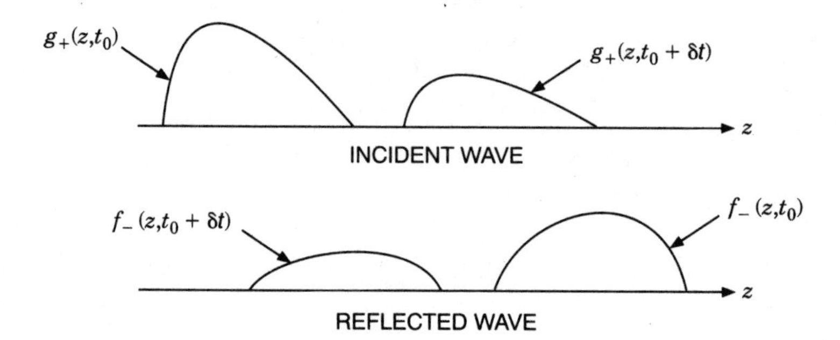

Figure 1.9 An illustration of propagation on a distortionless lossy line.

We use the above definition and (1.52) to rewrite the (1.70) and (1.71) in the form

$$g_+(z,t) = e^{-\alpha z} g(t - z/\nu) \tag{1.73}$$

$$f_-(z,t) = e^{+\alpha z} f(t + z/\nu) \tag{1.74}$$

The product αz is an exponent of the natural logarithm base e, hence it is dimensionless. It is assigned the dimensionless unit neper, so α is clearly given in nepers per meter. If we designate by α_{nep} the attenuation constant in nepers/meter, then for a given distance z the attenuation in a signal propagating from left to right is given by

$$\text{attenuation} = e^{-\alpha_{nep} z} \tag{1.75}$$

In manufacturers' specifications, attenuation constants are usually given in decibels per meter. If we designate by α_{dB} the attenuation constant in dB/m, then for a given distance z, the attenuation in a signal propagating from left to right is given by

$$\text{attenuation} = 10^{-(\alpha_{dB}/20)z} \tag{1.76}$$

We can convert between the two forms of attenuation constant if we observe that we must get equal results whichever way we compute attenuation. Equating expressions (1.73) and (1.74) and taking the natural logarithm of both sides we get

$$\alpha_{nep} = \frac{\ln(10)}{20}\alpha_{dB} \approx 0.115\alpha_{dB} \tag{1.77}$$

Distortionless lossy lines are very difficult to find in practice. Most lines have two conductors that are well insulated from each other, hence the leakage conductance G_0 is so negligibly small that it can be considered zero. The series resistance R_0 would also have to be so negligible as to be considered zero in order to satisfy (1.69). In that case we would be dealing with a lossless line. The purpose for discussing the distortionless line at this point was to familiarize the reader with the concept of attenuation and attenuation constants.

1.8 Conclusion

At this point it should be readily apparent that transmitting energy from point to point requires time, and that our ultimate concern is with the associated delays. Long transmission-lines will cause long delays, which in digital systems implies that clock frequencies will have to be reduced accordingly to make sure that transmitted signals are given the time needed to arrive at their destinations before any action is taken that they affect. The concepts introduced in this chapter will be used in later chapters to arrive at some useful conclusions on how to design digital systems.

Bibliography

1. W. C. Johnson, *Transmission Lines and Networks,* McGraw-Hill, New York, 1950.
2. P. C. Magnusson, *Transmission Lines and Wave Propagation,* Allyn and Bacon, Boston, Massachusetts, 1970.
3. S. R. Seshadri, *Fundamentals of Transmission Lines and Electromagnetic Fields,* Addison-Wesley, Reading, Massachusetts, 1971.

Problems

P1.1 Go through the detailed steps of obtaining (1.7) and (1.8) from (1.4) and (1.5).

P1.2 In connection with the traveling waves of (1.13) and (1.14),

(a) Show by substitution that (1.13) and (1.14) will satisfy (1.11) and (1.12).

(b) If $v(z,t) = g_+(z,t)$ of (1.13), then $i(z,t)$ can be related to $g_+(z,t)$ using (1.9) or (1.10). Carry out this procedure, thus verifying (1.34). Repeat this procedure in connection with (1.14).

P1.3 Apply the reasoning in (1.15) to (1.17) to $f_-(z,t)$ in (1.14) to arrive at figure 1.5 and the conclusion of (1.21).

P1.4 We would like to use a different method of establishing the speed of the traveling waves of (1.13) and (1.14).

(a) An observer sees a specific portion (phase) of the waveform

$$g_+(z,t) = g\left(t - \sqrt{L_0C_0}\, z\right)$$

and starts to run along the transmission line at just the right speed to keep abreast of this phase of the wave. Use this fact to determine the speed of the observer. (**Hint:** The observer is running at the speed needed to keep $g(\cdot)$ constant.)

(b) Repeat for the waveform

$$f_-(z,t) = f\left(t + \sqrt{L_0C_0}\, z\right)$$

P1.5 Carry out the substitution of (1.29) and (1.30) into (1.28), then substitute that into (1.9) to obtain (1.32).

P1.6 The transmission line shown in figure 1.10 has a steady-state incident voltage of 21 V and a steady-state reflected current of 0.2 A indicated.

(a) Determine $i_i(z,t)$ and $v_r(z,t)$ for the polarities indicated in the diagram.

(b) Determine the incident and the reflected power on the line. What is the power delivered to a load connected somewhere on the right end of the line?

P1.7 In example 1.3 expressions were given for the speed of propagation and the characteristic impedance of a coaxial cable.

(a) Find the ratio of D/d necessary to produce a cable with a characteristic impedance $Z_0 = 50\,\Omega$ if the dielectric is air.

(b) Find the speed of propagation if the cable dielectric is polyethylene which has a relative permittivity $\epsilon_r = 2.3$.

(c) What should be the ratio of D/d of the cable of part (b) to produce a characteristic impedance of $50\,\Omega$?

P1.8 Periodic time functions are averaged by integrating over one time period and dividing the result by the time duration of one period.

(a) Find the time average of $p_+(z,t)$ given in (1.61).

(b) Repeat for $p_-(z,t)$ given by (1.62).

P1.9 A lossless transmission line whose characteristic impedance is $50\,\Omega$ is operating in the sinusoidal steady state. The incident voltage wave has a magnitude of 10 V and the reflected current a magnitude of 0.15 A. Find the time averaged incident power, reflected power, as well as the power delivered to the right side of some arbitrary point z along the line.

P1.10 For the distortionless lossy line of section 1.7, show that (1.70) and (1.71) satisfy (1.7) and (1.8) provided that (1.69) holds.

P1.11 The characteristic impedance of a lossless line was found in section 1.4. Starting with (1.70), derive the expression for the characteristic impedance of a distortionless lossy line. Repeat for (1.71).

ANSWER: It is the same as that given in (1.33).

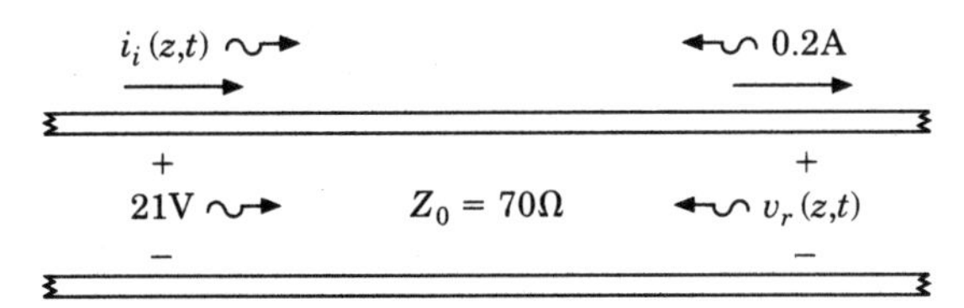

Figure 1.10 A lossless line with incident and reflected waves.

P1.12 A distortionless line has the parameters $C_0 = 70\,\text{pF/m}$, $L_0 = 0.35\,\mu\text{H/m}$, and $R_0 = 0.044\,\Omega/\text{m}$.

(a) What is the leakage conductance G_0 for this transmission line?
(b) Find the attenuation constant for this line in nepers and in dB. What is the attenuation in dB per 1000 meters?
(c) A 50 V sinusoidal signal is launched on the left side of the line and propagates to the right. What will be the value of the voltage when it has propagated 1000 meters to the right?
(d) For the previous part, find the signal power on the left end of the line and also determine the power level 1000 meters to the right of that point.

Chapter 2

Reflections on Transmission Lines

In the preceding chapter our discussion revolved around the concepts of incident and reflected waves, without any discussion of how the waves are established. It seemed that transmission lines were infinite in length and that the waves never encountered any terminations. In reality transmission lines do not continue endlessly in both directions. To be of practical use transmission lines are used to deliver signals or power from one location to another. There may be a source at one end of the line and a load at the other. Lines do not necessarily have to remain uniform. Two lines of different characteristic impedance can be connected at some point of discontinuity. Lines may split into two or more lines, or lumped impedances may be inserted anywhere along a line.

Kirchhoff's voltage and current laws have to be obeyed at all discontinuities. This determines the relationship between the incident and reflected waves, as will presently become apparent.

2.1 Wave Reflections on Terminated Lossless Lines

We will now depart from the concept of infinitely long lines and address ourselves to a finite lossless line, whose characteristic impedance is Z_0, of length l, with a power source at one end and a resistive termination at the other end, as seen in figure 2.1.

The line is initially deenergized so there is neither voltage nor current on the line. When the switch is closed a voltage wave of

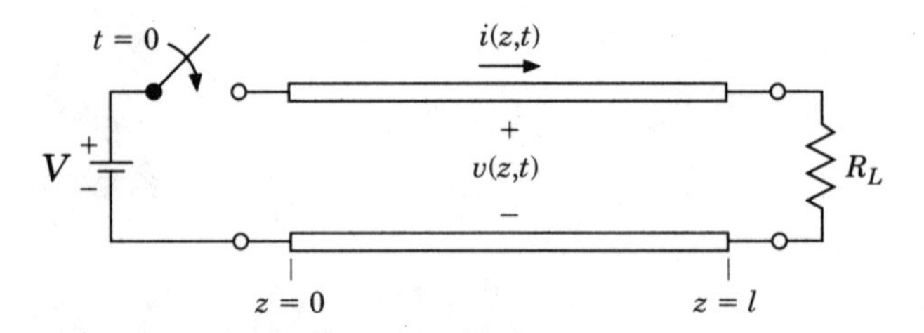

Figure 2.1 Lossless line with direct power source and resistive termination.

magnitude V starts propagating with speed ν from left to right on the line. We know from the previous chapter that the voltage wave is accompanied by a current wave V/Z_0. The two waves are shown in the top diagram of figure 2.3. At time

$$T = \frac{l}{\nu} \tag{2.1}$$

the waves reach the load resistor R_L at the right end of the line. The ratio of incident voltage to incident current is Z_0, and this does not satisfy Ohm's law for the load resistance R_L. At this point a reflected wave is launched so that Ohm's law requirement at the load will be satisfied.

The voltage and current conventions shown in figure 1.6 will be used. We will start our analysis by recalling from (1.45) and (1.46) that the voltage and current on the line are composed of incident and reflected components

$$v(z,t) = v_+(z,t) + v_-(z,t) \tag{2.2}$$

$$i(z,t) = i_+(z,t) + i_-(z,t) \tag{2.3}$$

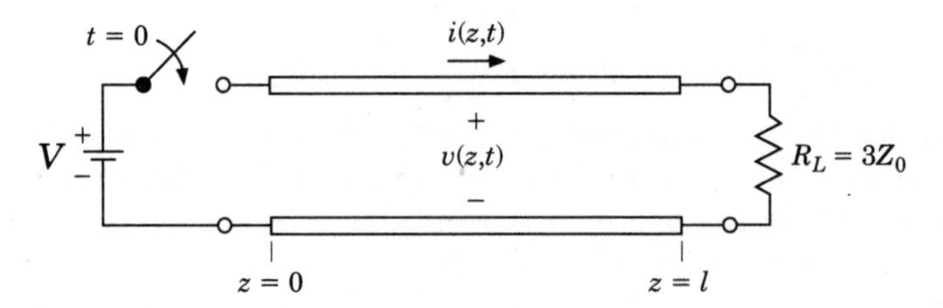

Figure 2.2 Lossless line with direct power source and resistive termination of $3Z_0$.

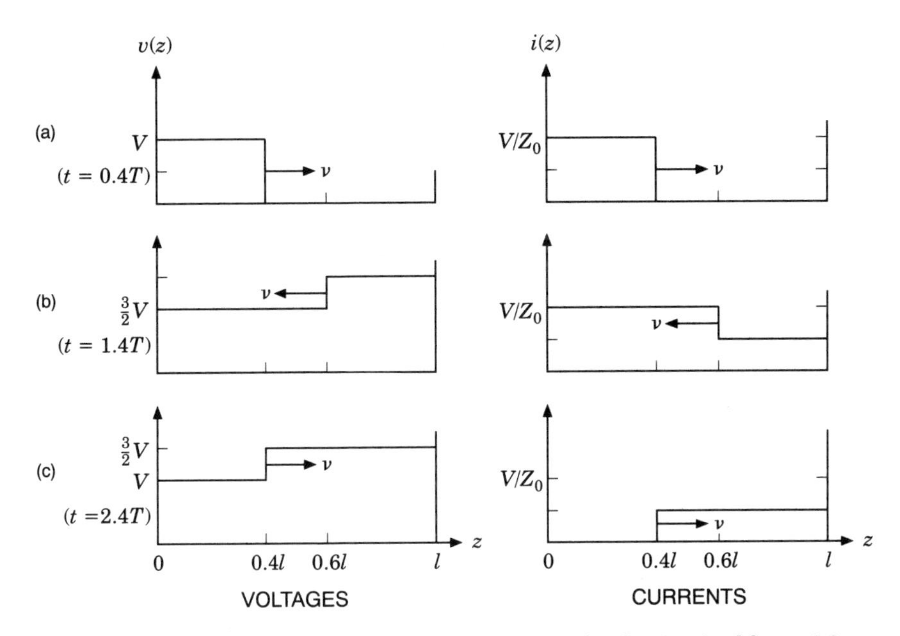

Figure 2.3 Voltage and current distributions on the line for the circuit of figure 2.2 at various times.

We also recall from (1.49) and (1.50) that voltage and current components are related by

$$i_+(z,t) = \frac{1}{Z_0} v_+(z,t) \tag{2.4}$$

$$i_-(z,t) = -\frac{1}{Z_0} v_-(z,t) \tag{2.5}$$

At the load, the ratio of (2.2) to (2.3) has to equal R_L. We therefore substitute (2.4) and (2.5) into (2.3), then take the ratio of (2.2) to (2.3) and set that equal to R_L. From that we determine the ratio of $v_-(l,T)$ to $v_+(l,T)$. The result is ρ_{Lv}, the *voltage reflection-coefficient* at the load, which is given by

$$\rho_{Lv} \equiv \frac{v_-(l,T)}{v_+(l,T)} = \frac{R_L - Z_0}{R_L + Z_0} \tag{2.6}$$

Substituting (2.4) and (2.5) into (2.6) we solve for the ratio of $i_-(l,T)$ to $i_+(l,T)$. The result is ρ_{Li}, the *current reflection-coefficient* at the load, expressed by

$$\rho_{Li} \equiv \frac{i_-(l,T)}{i_+(l,T)} = -\frac{R_L - Z_0}{R_L + Z_0} \tag{2.7}$$

For convenience we will henceforth use only one reflection coefficient expressed by

$$\rho_L \equiv \rho_{Lv} = -\rho_{Li} = \frac{R_L - Z_0}{R_L + Z_0} \tag{2.8}$$

The above definition corresponds to the voltage reflection coefficient of (2.6). We have to be conscious, however, that *the reflection coefficient for current is the negative of the reflection coefficient for voltage.* It will be found that (2.8) is an extremely useful expression for solving transmission-line problems. Some examples will illustrate its use.

Example 2.1 Consider the transmission line of figure 2.2. When the switch is closed at time zero, a voltage wave of value V is launched, as seen in the left diagram of figure 2.3a, so that

$$v_+(z,t) = V, \qquad \text{for} \quad 0 \le t \tag{2.9}$$

This voltage wave is accompanied by a current wave, as shown in the right diagram of figure 2.3a, whose value is found by applying (2.4) to (2.9)

$$i_+(z,t) = \frac{V}{Z_0}, \qquad \text{for} \quad 0 \le t \tag{2.10}$$

Both of those waves propagate from left to right at speed ν, and once they reach the right side of the line the voltage and current that they have created stay on the line permanently. Superposition holds, so that any additional voltage or current waves that are subsequently created keep adding to those that are already there.

When the incident wave arrives at the load at time T, a reflected wave is required in order to satisfy Ohm's law at the load. We find using (2.8) that the *voltage reflection coefficient* at the load has a value

$$\rho_L = \frac{3Z_0 - Z_0}{3Z_0 + Z_0} = \frac{1}{2} \tag{2.11}$$

We multiply (2.9) by ρ_L to find

$$v_-(z,t) = \frac{V}{2}, \qquad \text{for} \quad T \le t \tag{2.12}$$

Since the reflection coefficient for current is the negative of the reflection coefficient for voltage, we multiply (2.10) by $-\rho_L$ to find

$$i_-(z,t) = -\frac{V}{2Z_0}, \qquad \text{for} \quad T \le t \tag{2.13}$$

We could have avoided the use of $-\rho_L$ to obtain the above result by simply applying (2.5) to (2.12) to obtain (2.13).

When (2.12) is added to (2.9) we find the total load voltage is $3V/2$, and when (2.13) is added to (2.10) we see that the total load current is $V/(2Z_0)$. These have the ratio $3Z_0$, satisfying the Ohm's law requirement at the load. As the reflected waves propagate in the direction of the left-hand source, they establish the voltage of $3V/2$ and the current $V/(2Z_0)$ on the entire line, as shown in figure 2.3b.

We will now analyze the events at the source end of the transmission line. When the reflected wave arrives at the source end it sees a source impedance R_S of zero ohms. At the source (2.8) takes the form

$$\rho_S = \frac{R_S - Z_0}{R_S + Z_0} = \frac{0 - Z_0}{0 + Z_0} = -1 \tag{2.14}$$

We multiply (2.12) by ρ_S to obtain the second incident voltage wave

$$v_+(z,t) = -\frac{V}{2}, \qquad \text{for} \quad 2T \le t \tag{2.15}$$

and we multiply (2.13) by $-\rho_S$ to obtain the second incident current wave

$$i_+(z,t) = -\frac{V}{2Z_0}, \qquad \text{for} \quad 2T \le t \tag{2.16}$$

The effects of these last two waves are indicated in figure 2.3c. It is noteworthy that we are using *superposition* to solve this problem. It is the superposition of all the waves that determines the voltage and current values on the line at all times. ■

2.2 Reflection or Lattice Diagrams

The solution for the circuit of figure 2.2 presented above is rather tedious. It is made much easier through the use of a *reflection diagram* which is also referred to as a *lattice diagram*. An example of a lattice diagram for the problem of example 2.1 is given in figure 2.4.

The lattice diagram provides a convenient graphical method for presenting the data associated with the calculations involved with successive reflections. The diagram has a horizontal z-axis and a vertical (downwardly increasing) t-axis. The locus of a wave traveling in the increasing z-direction is a line sloping downward from left to right, whereas the locus of a wave traveling in the decreasing z-direction is a line sloping downward from right to left. Each locus is labeled with the value of the wave. The magnitude of each reflection is obtained by multiplying the magnitude of the preceding wave by the reflection coefficient at the point where the reflection takes place.

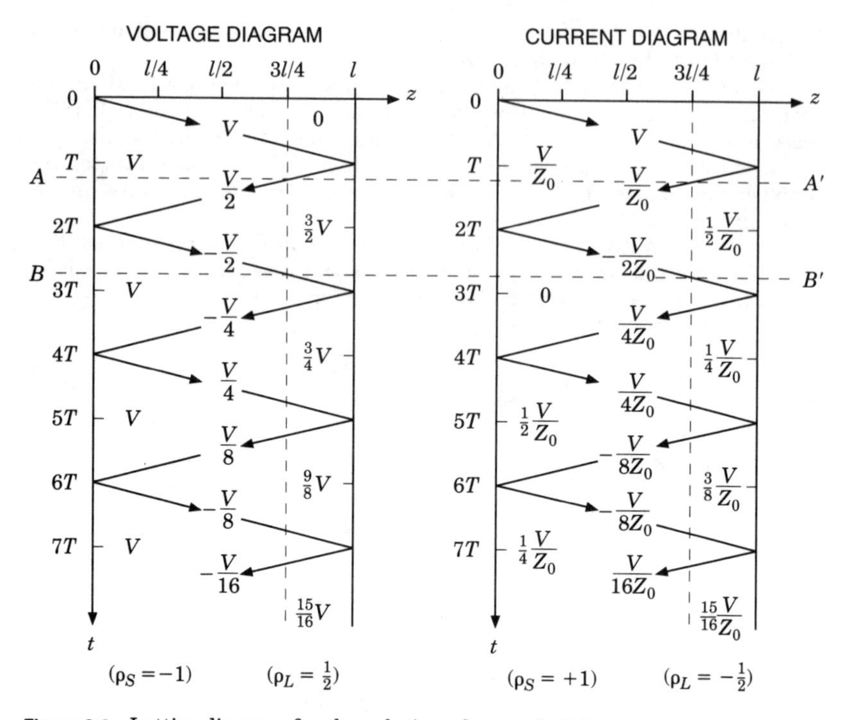

Figure 2.4 Lattice diagram for the solution of example 2.1.

Inside each triangular area is the sum of the waves above that region of the chart.

When drawing the diagram for current it must be kept in mind that the reflection coefficients for current are of opposite sign of those for voltage. Alternately, the lattice diagram for current can be obtained from the lattice diagram for voltage using (2.4) and (2.5) to determine the values of the currents for the diagonal traces. The sums of the currents, however, are not directly related to the sums of voltages.

The lattice diagram for voltage (or current) can be used to deduce the voltage (or current) at any point on the line as a function of time. The vertical dashed lines in the lattice diagrams represent the point $z = \frac{3}{4}l$ on the transmission line. It is only necessary to observe the voltages (or currents) marked inside the triangles to determine the time function for that variable. On the voltage lattice-diagram, the dashed line passes through the 0 volt triangle during $0 \le t < \frac{3}{4}T$. When $\frac{3}{4}T \le t < \frac{5}{4}T$ it passes through the first V volt triangle. Similarly for the other regions. The voltage and current waveforms in figure 2.5 were based on the lattice diagrams of figure 2.4.

We will postpone a formal discussion of the steady-state behavior of transmission lines to a later section. For the moment we will rely

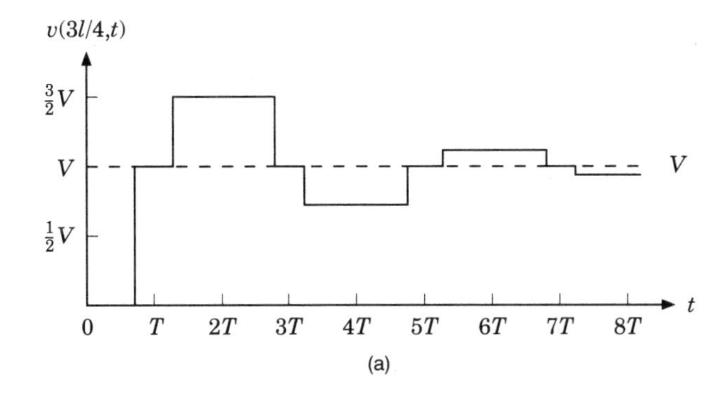

(a)

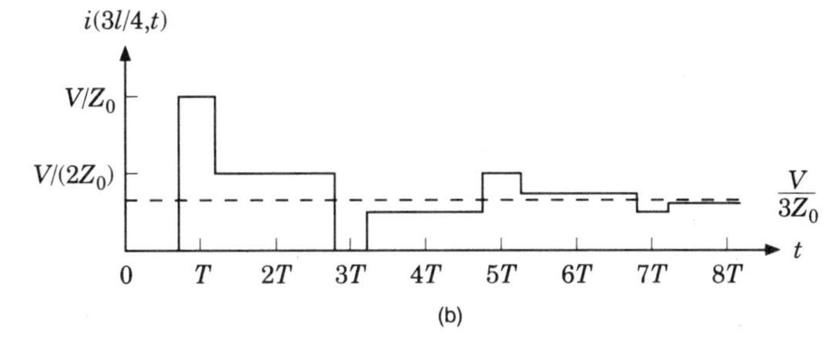

(b)

Figure 2.5 Voltage waveform (a) and current waveform (b) at $3 = \frac{3}{4}l$. Steady-state values are indicated with dashed lines.

on an intuitive approach. We inspect the circuit shown in figure 2.2 and conclude, without too much difficulty, that in the steady state the lossless transmission-line, which is composed of differential series-inductance and differential shunt-capacitance, behaves as an ordinary pair of wires. Hence the voltage on the line will be the battery voltage V and the steady-state current will be the battery voltage divided by the load resistance, taking the value $V/(3Z_0)$. It is seen in figure 2.5 that the voltage and current on the line oscillate around the steady-state values.

Lattice diagrams have other uses. They can also be used to find the voltage and current distributions on the line at any instant of time. The horizontal dashed line AA' was used for determining the voltage and current distribution on the line at time $t = \frac{5}{4}T$. On the voltage lattice-diagram the dashed line AA' passes through the V volt triangle for $0 \leq z < \frac{3}{4}l$ and through the $\frac{3}{2}V$ volt triangle for $\frac{3}{4}l \leq z \leq l$. It intersects the reflected wave $V/2$, so we also have the information needed to label the direction of the wavefront, as shown in the top left graph of figure 2.6. The same technique was used to obtain the current distribution at $t = \frac{5}{4}T$ as well as the distributions for $t = \frac{11}{4}T$.

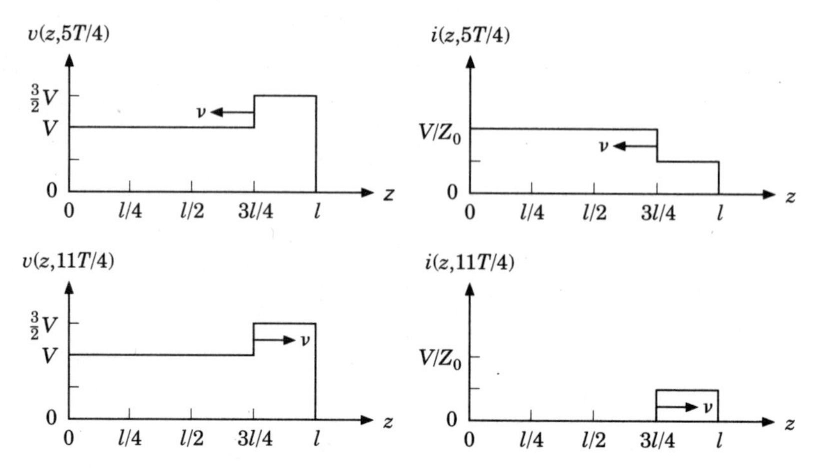

Figure 2.6 Voltage and current distributions as a function of z for $t = \frac{5}{4}T$ and $t = \frac{11}{4}T$.

2.3 Effect of Source Resistance

The addition of a source-resistance only complicates matters a trifle. Consider the circuit in figure 2.7.

The reflection coefficients at the source and the load ends of the line are given by

$$\rho_S = \frac{R_S - Z_0}{R_S + Z_0} \tag{2.17}$$

$$\rho_L = \frac{R_L - Z_0}{R_L + Z_0} \tag{2.18}$$

When the switch in figure 2.7 is closed, the voltage V is applied through an output impedance R_S to the transmission line whose characteristic impedance is Z_0. We have voltage division at the input of the line, and the first incident wave V'_+ has a value

$$V'_+ = \frac{Z_0}{R_S + Z_0} V \tag{2.19}$$

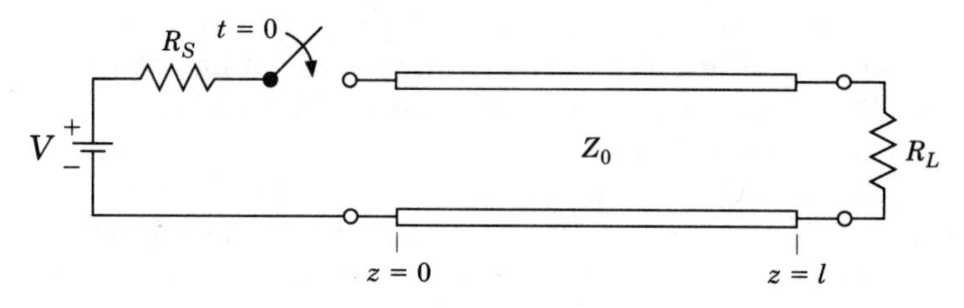

Figure 2.7 Lossless line with direct power source, and resistive source and load terminations.

From this point on the analysis proceeds in a very straightforward manner, as illustrated below.

Example 2.2 Consider the transmission line of figure 2.7. The source resistance R_S has a value of Z_0 and the load resistance R_L is an open circuit. When the switch is closed, we find using voltage division that the first incident wave is $V'_+ = V/2$. At time T this voltage wave arrives at the load and encounters a reflection coefficient of unity so it is fully reflected. When the reflected voltage arrives at the source, where the reflection coefficient is zero, there are no further reflections.

We see that the voltage on the entire line is $V/2$ at time T. At time $2T$ it takes on a value of V. There are no further reflections so the line is now in its steady state. ■

Example 2.3 Consider the transmission line of figure 2.7. The source resistance R_S has a value of Z_0 and the load resistance R_L is a short circuit. When the switch is closed, we find using voltage division that the first incident wave is

$$V'_+ = \frac{V}{2}$$

At time T this voltage wave arrives at the load and encounters a reflection coefficient $\rho_L = -1$. The incident wave changes sign on reflection. When the reflected wave arrives at the source, where the reflection coefficient is zero, there are no further reflections.

We see that the voltage on the entire line is $V/2$ at time T. At time $2T$ it takes on a value of zero which is the steady-state value of voltage on this line that has a short circuit for a load. ■

2.4 The Steady State on Lossless Lines

In section 2.3 we relied on an intuitive approach to find the steady-state voltage and current on the transmission line. It seemed reasonable that a lossless line consisting entirely of distributed series inductance and distributed parallel capacitance should act like a pair of ordinary wires in the steady state. We will see through the proof which follows that our intuition has served us well.

As was done in (2.19), the initial voltage wave launched by the source will be designated V'_+. When this voltage arrives at the load, a reflection takes place. The returning wave has the value $\rho_L V'_+$. At the source end another reflection occurs. The forward wave due to this last reflection is $\rho_S \rho_L V'_+$. We can account for any number of reflections by this method. These repeated reflections are demonstrated schematically in figure 2.8. The total voltage on the line is the superposition of all the reflected voltages. It is given by the sum of infinite terms

$$v(z,t) = V'_+\left[1 + \rho_L + \rho_L\rho_S + \rho_L^2\rho_S + \rho_L^2\rho_S^2 + \rho_L^3\rho_S^2 + \cdots\right] \qquad (2.20)$$

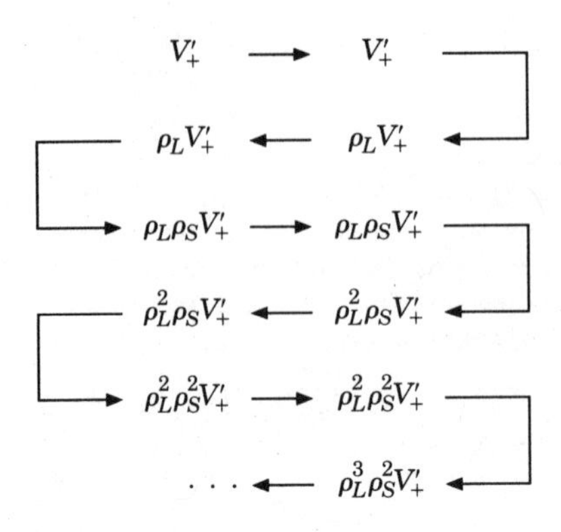

Figure 2.8 A lattice-like diagram for the circuit of figure 2.7.

The above expression can be used to find the voltage at any point on the line for a specific time by first establishing how many terms need to be summed. Referring to a lattice diagram can be very helpful to determine how many reflections have to be considered for a given elapsed time from switch closure.

For the steady state we need to sum an infinite number of terms in (2.20). We do this by factoring (2.20) into the form

$$v(z,t) = V'_+(1 + \rho_L)\left[1 + \rho_S\rho_L + \rho_S^2\rho_L^2 + \cdots\right] \tag{2.21}$$

The term in the brackets is the sum of an infinite geometric series with a common ratio $\rho_S\rho_L$. The sum for an infinite geometric series is well known and, if the common ratio is less than unity in magnitude, can be written in the closed form

$$1 + x + x^2 + x^3 + x^4 + \cdots = \frac{1}{1 - x}, \qquad \text{for} \quad |x| < 1 \tag{2.22}$$

Applying (2.22) to (2.21) we find the closed form expression for the steady-state voltage

$$V_{ss} = V'_+\frac{1 + \rho_L}{1 - \rho_S\rho_L}, \qquad \text{for} \quad |\rho_S\rho_L| < 1 \tag{2.23}$$

To find the steady-state expression for current we note that the initial current wave is V'_+/Z_0 and also that the reflection coefficients for current are the negatives of the reflection coefficients for voltage.

Using these two observations in (2.23) we find the expression for steady-state current

$$I_{ss} = \frac{V'_+}{Z_0}\frac{1 - \rho_L}{1 - \rho_S\rho_L}, \qquad \text{for} \quad |\rho_S\rho_L| < 1 \tag{2.24}$$

We can use (2.23) and (2.24) to verify the steady-state values indicated in figure 2.5. As an alternative we can substitute into the last two expressions for V'_+, ρ_L, and ρ_S from (2.17), (2.18), and (2.19) to obtain the equations

$$V_{ss} = \frac{R_L}{R_S + R_L}V \tag{2.25}$$

$$I_{ss} = \frac{V}{R_S + R_L} \tag{2.26}$$

Examining the last two equations while inspecting figure 2.7, it is readily concluded that in the steady state the lossless transmission-line behaves like a pair of ordinary wires. It is as if the source, with its output resistance R_S, is directly connected to the load R_L.

2.5 Non-constant Source Excitations — Precursors

There are times when transmission line problems are encountered in which non-constant sources are connected at the input. The most straightforward method of solving these problems is by noting landmarks on the waveshape of the exciting source and additionally using precursors*. The easiest way to explain this method is through an example.

Example 2.4 Consider the transmission-line problem of figure 2.9 which has the non-constant source $v_S(t)$ connected on the left. It has a trapezoidal waveshape starting with a leading edge of 1024/3 volts and dropping linearly toward 512/3 volts at its trailing edge at time $T/2$. In this particular case the landmarks are the leading and trailing edges of the input pulse. Once the analysis is performed for the leading and trailing edges, the solution for the intermediate times will become obvious.

*Precursor — Forerunner which announces the arrival of something.

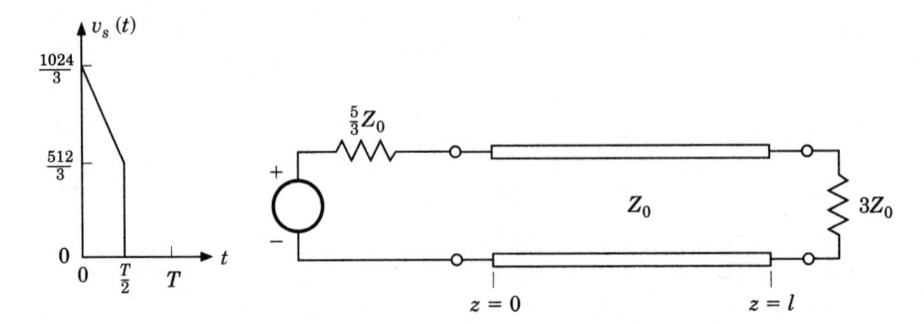

Figure 2.9 Lossless line with pulse power source and resistive terminations.

The reflection coefficients at the ends of the line are $\rho_S = \frac{1}{4}$, and $\rho_L = \frac{1}{2}$. At time $t = 0$, the leading edge of the applied voltage, which has an amplitude of 1024/3 volts, gives rise to an incident voltage wave on the line of 128 volts. At time $T/2$ the trailing edge of the pulse, which is 512/3 volts, creates an incident voltage wave of 64 volts. We proceed to draw a two-track lattice-diagram, one track for the leading edge of the applied wave and one track for the trailing edge, as shown in figure 2.10. The pulse does not exist outside the shaded region of the lattice diagram.

Dashed vertical lines are drawn at $z = l/4$ and at $z = 7l/8$ in an effort to find the time dependence of the voltage at those locations. The dashed line at $z = l/4$ passes through the 128 volt track at $t = T/4$ and then through the 64 volt track at $t = 3T/4$. We can use this to draw the first pulse in figure 2.11a. The dashed line then passes through the 64 volt, 32 volt, 16 volt, and 8 volt tracks. The result

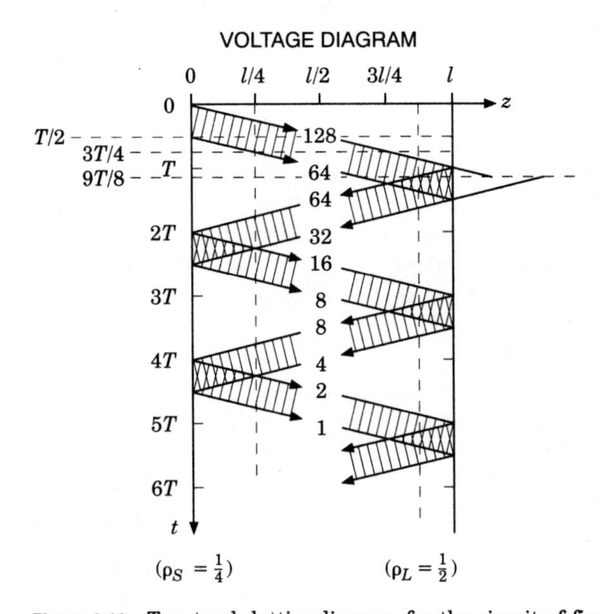

Figure 2.10 Two-track lattice-diagram for the circuit of figure 2.9.

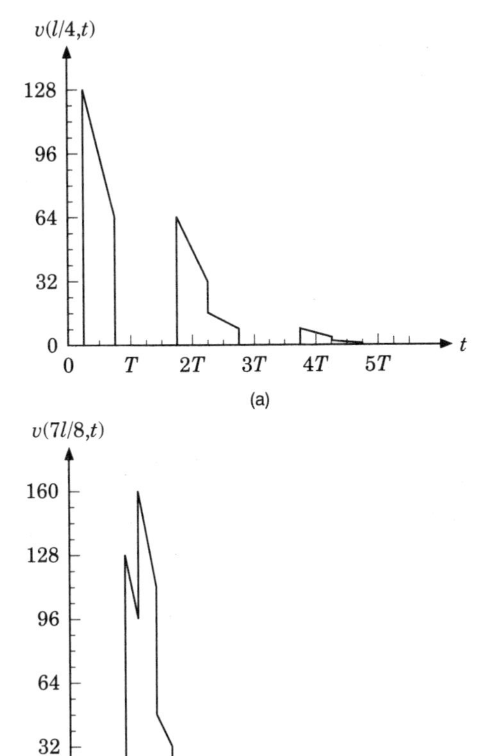

Figure 2.11 Voltage waveforms at (a) $z = l/4$ and (b) $z = 7l/8$.

of this is shown in the second, somewhat more complicated pulse in figure 2.11a. This process can be carried out for any desired value of time.

The dashed line at $z = 7l/8$ in figure 2.10 creates a somewhat more complicated time dependent waveform because the reflected pulse overlaps with the incident pulse. The voltages of the two pulses have to be summed to obtain the resultant waveforms shown in figure 2.11b.

The lattice diagram of figure 2.10 has horizontal dashed lines drawn at $t = T/2$, $t = 3T/4$, and at $t = 9T/8$. These lines let us find the voltage distribution on the line at these instants of time. The results are shown in figure 2.12. It is noteworthy in figure 2.12a that the voltage pulse appears reversed in relation to the one shown in figure 2.9, so that the leading edge of the pulse is at $z = l/2$ while the trailing edge is at $z = 0$. This makes perfectly good sense since the leading-edge voltage makes its presence felt first at the input to the line. The trailing edge voltage does not make its presence felt for another $T/2$ seconds. Figure 2.12b shows the voltage distribution at $t = 3T/4$.

Figure 2.12c shows the voltage distribution at $t = 9T/8$. The complications involved in drawing this figure require some elaboration. In this diagram the incident pulse is depicted sliding toward the end of the line, while the reflected pulse is shown sliding in from beyond the end of the line. The line beyond $z = l$ is fictitious, but it is added to the diagram as an aid in the analysis.

The propriety of drawing the reflected pulse sliding in from beyond the physical end of the line, which is the principle of precursors, can be justified quite readily from an examination of figure 2.10. There we have extended to the right the dashed line at $t = 9T/8$. It is readily apparent that this line intersects the 128 V value of the incident pulse at the fictitious point $z = 9l/8$ and that it intersects the 64 V value of the incident pulse at $z = 5l/8$. The incident pulse is shown in figure 2.12c propagating to the right at a speed ν. Further examination of figure 2.10 shows that the dashed line intersects the reflected pulse voltage of 32 V at $z = 11l/8$ and the value of 16 V at $z = 7l/8$. The reflected pulse is shown in figure 2.12c propagating to the left at a speed ν.

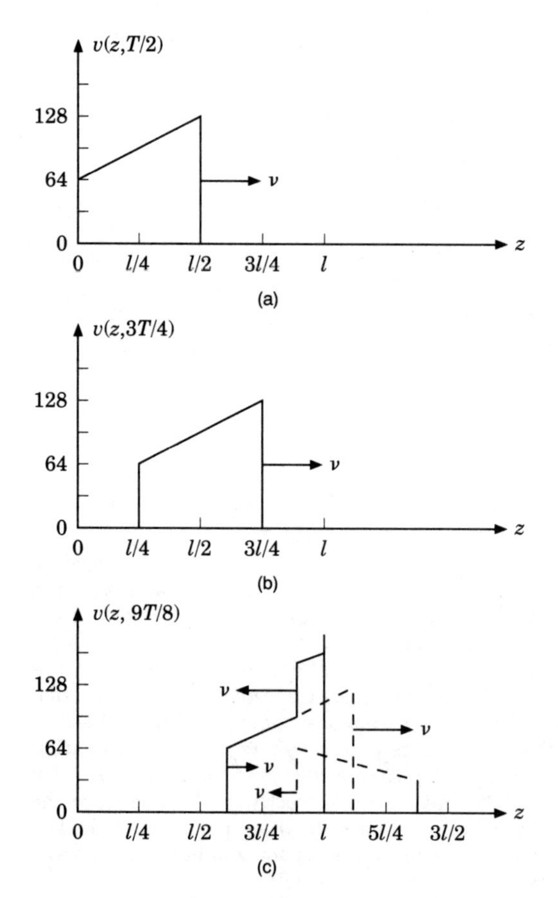

Figure 2.12 Voltage distribution on the line at (a) $t = T/2$, (b) $t = 3T/4$, and (c) $t = 9T/8$.

In the above analysis the reflected wave is being treated as equivalent to a wave starting life at $t = 0$ at the fictitious point on the line, $z = 2l$. Consequently figure 2.12c shows two waves meeting at the load: the incident wave which started life at $t = 0$ and $z = 0$, and a reflected wave (one half in height in this case), the *precursor*, which started life at the same time but at $z = 2l$. ■

The concept of using precursors along a fictitious part of the line can be extended to reflections at the load which occur at other times. For the reflection at $t = 3T$ at $z = l$ we can use a precursor starting T seconds earlier at $t = 2T$ at $z = 2l$. For a reflection at time $t = 5T$ at location $z = l$ we can use a precursor starting at $t = 4T$ at $z = 2l$.

2.6 Series and Parallel Terminated Lines

There are a number of different ways of terminating transmission lines to prevent undesired reflections. In digital circuits we think of signals originating at the output of digital devices which, for the sake of simplicity and without any loss of generality, we will model as digital-logic gates. These signals then propagate to the input terminals of other digital devices, which will also be modeled as digital-logic gates. The output impedance of high speed digital logic devices is fairly low, usually less than 15 Ω. The input impedance of high speed logic devices is very high, usually a few kilo-ohms. The characteristic impedances of printed circuit board traces, which connect the digital logic devices, range from as low as 25 Ω to a few hundred ohms. In this section we will therefore simplify the models by considering the signal as originating from an ideal voltage source, and propagating on a transmission line toward a load possessing infinite input impedance. In later sections and chapters we will take into consideration the details of the impedances loading the line.

The two principal methods of terminating such systems are shown in figure 2.13. When a line is terminated at either end with a resistor equal to its characteristic impedance, then the line is said to be *matched*. If there are no other overriding considerations, then the series terminating source resistance R_S and the parallel terminating load resistance R_L are chosen to match the line. There are tradeoffs in both systems of termination, which we will discuss by assuming that the sending gate in figure 2.13 transmits a step voltage of amplitude V.

The series terminated system shown in figure 2.13a has the advantage that in the high-voltage steady-state, the terminating resistance R_S dissipates no power. The disadvantage is that if there are gate inputs connected along the length of the transmission line, then a gate connected at $z = 0$ must wait till $t = 2T$ to get the full voltage V

at its input. There is a progressively shorter delay for gates connected closer to the output end of the line.

The parallel terminated system shown in figure 2.13b has the disadvantage that in the high-voltage steady-state, the terminating resistance R_L dissipates V^2/R_L watts of power. The advantage is that if there are gate inputs connected along the length of the transmission line, then a gate connected at $z = 0$ gets the full voltage V immediately, whereas a gate input connected at $z = l$ needs to wait only till $t = T$ to get the full voltage V at its input. Accordingly the parallel terminated system is faster than the series terminated system for gate inputs connected near the sending end of the transmission line.

2.7 Cascaded Lines — Transmission Coefficients

A non-uniform line is one in which the geometry changes as a function of the distance z. The analysis of transmission lines with non-uniform geometries can be very complicated. We will confine ourselves to the simple case of a non-uniformity created by the cascade connection of two lines as shown in figure 2.14.

We will use superposition to divide the solution of this problem into two parts. We will first assume that the incident wave v_{r2} on the right side of the interface is zero and determine the consequences of the wave v_{i1} which is incident on the left side of the interface. This wave gives rise to the reflected wave v_{r1} and to the transmitted wave v_{i2} which propagates away from the right side of the interface. In the second part of the solution we will assume that the incident

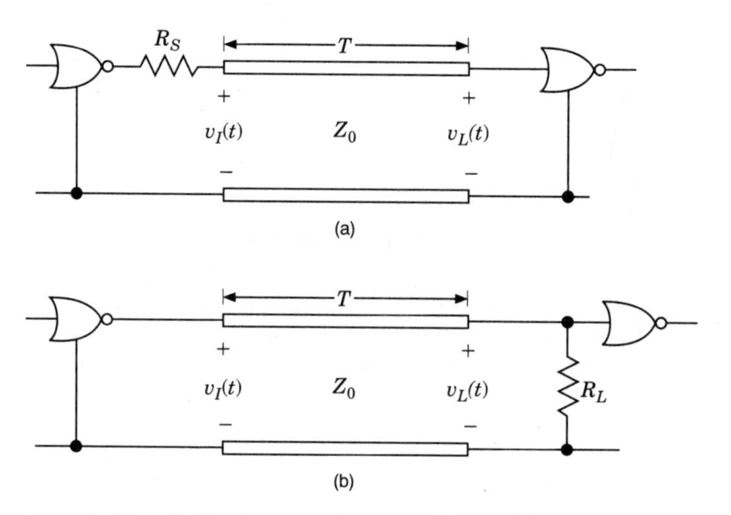

Figure 2.13 (a) Series terminated system, (b) parallel terminated system.

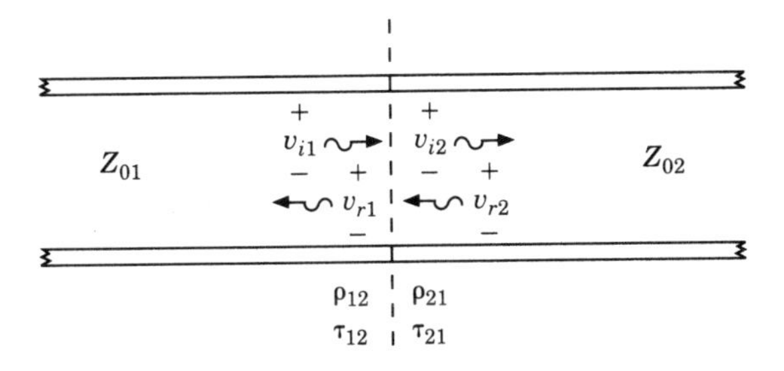

Figure 2.14 Cascaded transmission-lines.

wave v_{i1} on the left side of the interface is zero and determine the consequences of the wave v_{r2} which is incident on the right side of the interface. This wave gives rise to the reflected voltage v_{i2} and to the transmitted voltage wave v_{r1} which propagates away from the left side of the interface.

The incident voltage v_{i1} arriving at the interface from the left perceives the right-hand transmission-line as a load of impedance Z_{02}. Therefore ρ_{12}, the reflection coefficient which will be used to calculate the reflected wave v_{r1}, is given by

$$\rho_{12} = \frac{Z_{02} - Z_{01}}{Z_{02} + Z_{01}} \tag{2.27}$$

Whereas the reflection coefficient is used to determine the reflected wave at an interface, the *transmission coefficient* is used to determine the wave transmitted to the other side of an interface. For the interface illustrated in figure 2.14 the transmission coefficient τ_{12} is used to find the transmitted wave v_{i2} from the incident wave v_{i1}, and is defined in terms of the ratio

$$\tau_{12} \equiv \frac{v_{i2}}{v_{i1}} \tag{2.28}$$

To determine τ_{12}, we observe that the voltages on the left and right side of the interface must be equal. Again it is noted that superposition is being used, hence the incident voltage v_{r2} on the right side of the interface is zero. After the incident wave arrives we have v_{i1} and v_{r1} on the left of the interface, whose sum must equal the voltage on the right side of the interface which is v_{i2}, hence

$$v_{i1} + v_{r1} = v_{i2} \tag{2.29}$$

which can be restated as

$$(1 + \rho_{12})v_{i1} = \tau_{12}v_{i1} \tag{2.30}$$

leading to the final result

$$\tau_{12} = 1 + \rho_{12} \tag{2.31}$$

Using (2.27) we can get an independent expression for the transmission coefficient in the form

$$\tau_{12} = \frac{2Z_{02}}{Z_{02} + Z_{01}} \tag{2.32}$$

We can extend the results that were obtained for the left transmission-line to apply to the right one by expediently changing subscripts in (2.27), (2.31), and (2.32) to obtain

$$\rho_{21} = \frac{Z_{01} - Z_{02}}{Z_{01} + Z_{02}} \tag{2.33}$$

$$\tau_{21} = 1 + \rho_{21} \tag{2.34}$$

$$\tau_{21} = \frac{2Z_{01}}{Z_{01} + Z_{02}} \tag{2.35}$$

When (2.27) is compared with (2.33), it is noted that at an interface $\rho_{21} = -\rho_{12}$.

Example 2.5 Consider the two-line problem shown in figure 2.15. We wish to draw a lattice diagram for this circuit.

The reflection and transmission coefficients shown in figure 2.15 were obtained using (2.27), (2.31), (2.33), and (2.34). The lattice diagram is shown in figure 2.16. The right section of the diagram was drawn with a horizontal axis that is 1.5 times that of the left section. This was done for the sake of convenience. Otherwise the angle at which the rightmost diagonal lines are drawn would not match that of the diagonal lines in the left section of the lattice diagram, causing a great deal of inconvenience.

The step voltage of the source on the left creates an incident wave of 625 volts. At the interface we get, by using ρ_{12}, a reflected wave of 125 volts. The transmitted wave of 750 volts was obtained using the transmission coefficient τ_{12}. The process is then repeated to continue with the rest of the lattice diagram. ■

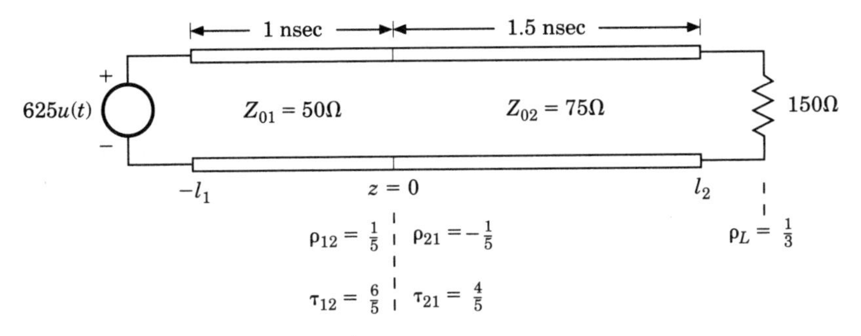

Figure 2.15 Cascade connected lines.

Example 2.6 In figure 2.17a we have a signaling gate, represented for convenience by the step source $81u(t)$, connected to one receiving gate at AA' and one at BB'. The receiving gates have a very high input impedance which for our purposes can be considered infinite. The properly parallel terminated connection has a resistor of value Z_0 placed at the extreme right of the transmission line. In this way the step voltage of 81 V arrives at the rightmost gate after $2T$ seconds, and there are no subsequent reflections. We wish to determine the behavior of the circuit if the terminating resistance is misplaced, as shown in figure 2.17b.

To assess the consequences of the error, we start by noting that a signal arriving from either side toward the interface AA' sees a parallel combination of the transmission-line characteristic-impedance Z_0 in parallel with the misplaced

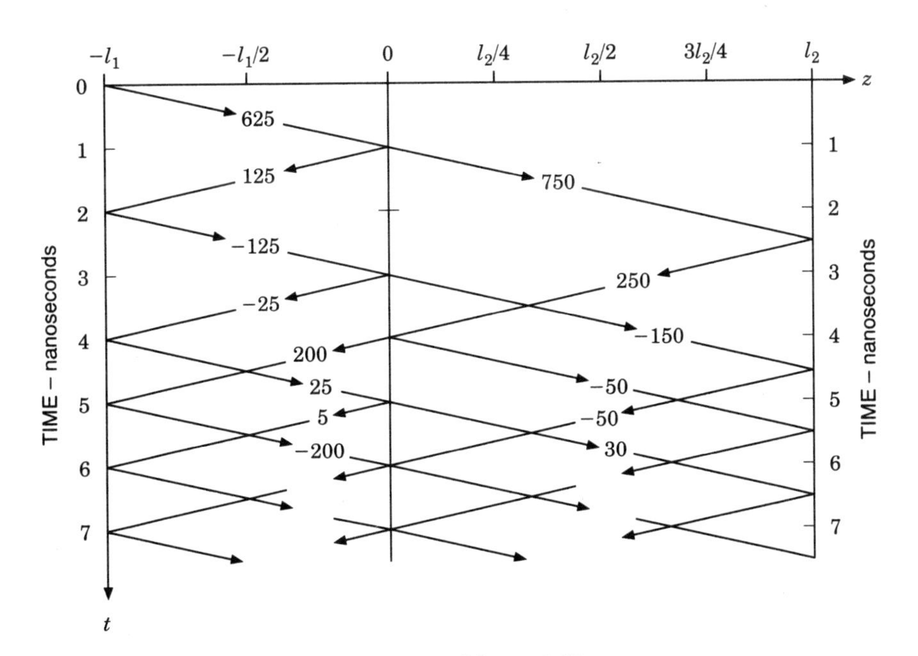

Figure 2.16 Lattice diagram for the circuit of figure 2.15.

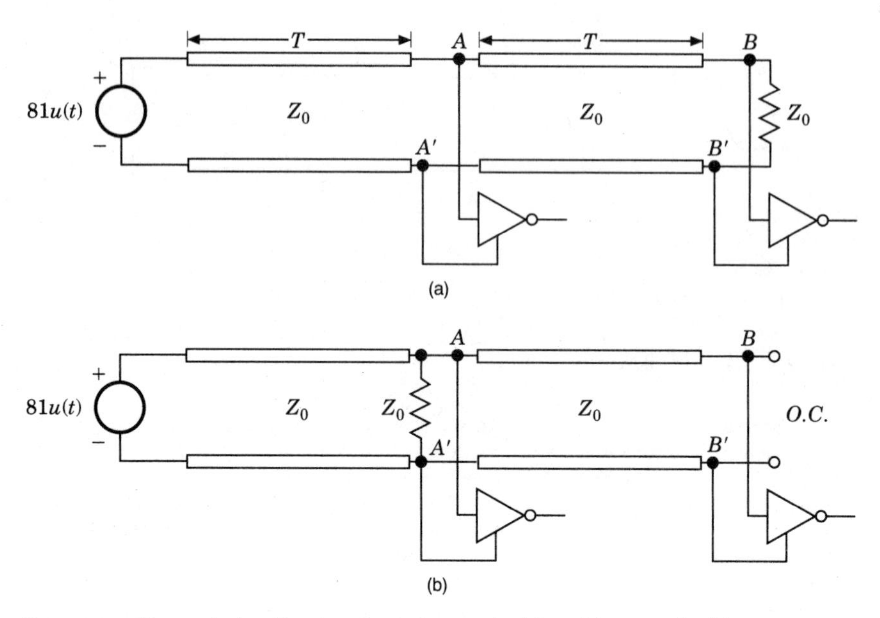

Figure 2.17 Transmission line terminated properly (a) and improperly (b).

terminating resistor Z_0. Hence the reflection coefficient at AA', for signals traveling in either direction, is

$$\rho_{AA'} = \frac{Z_0/2 - Z_0}{Z_0/2 + Z_0} = -\frac{1}{3} \tag{2.36}$$

and as a consequence

$$\tau_{AA'} = 1 + \rho_{AA'} = \frac{2}{3} \tag{2.37}$$

The reflection coefficient at the left-hand sending end is -1 and at the extreme right receiving end it is $+1$. With this information in hand we can now draw the lattice diagram shown in figure 2.18. It is necessary to be aware of the fact that on the lattice diagram at the interface AA' both reflection and transmission have to be considered at the same time. Thus for example at $t = 3T$ the 54 V signal causes a reflection on the right side of -18 V while the 27 V signal causes a transmission of $+18$ V to the right side of the line. Hence the total signal transmitted is zero volts.

The steady-state voltage on this line is 81 V. The highest value attained is 108 V which represents a very large 33% overshoot. This is illustrated in the voltage waveforms shown in figure 2.19. The gate at BB' reaches 84 V, which is less than 4% from the steady-state value of 81 V after $10T$ seconds. This is a delay five times as great as that for the correctly terminated system of figure 2.17a. ■

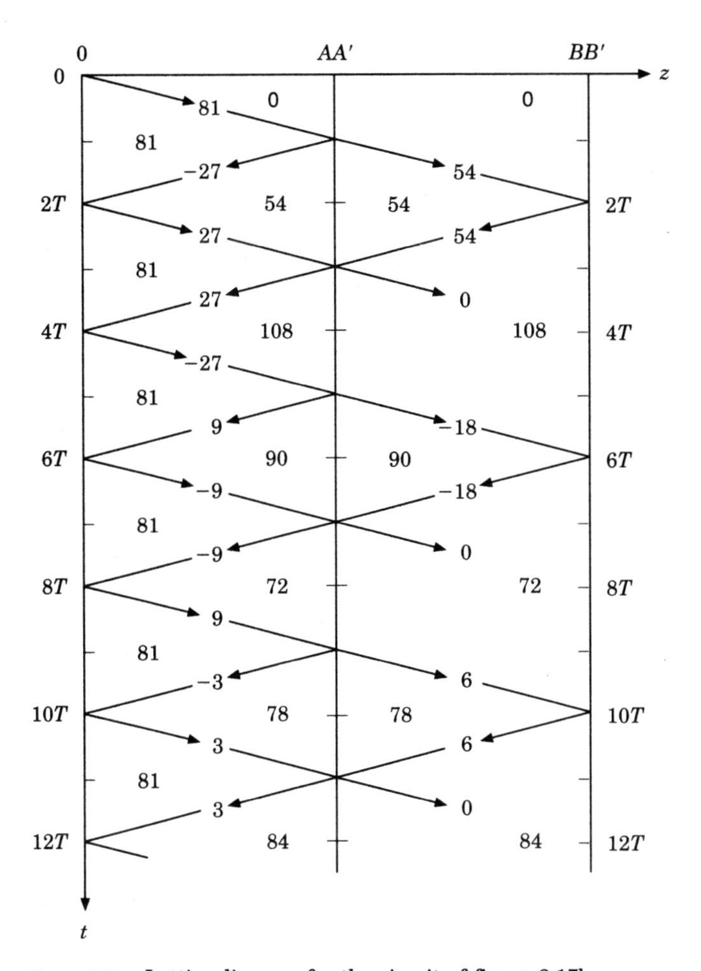

Figure 2.18 Lattice diagram for the circuit of figure 2.17b.

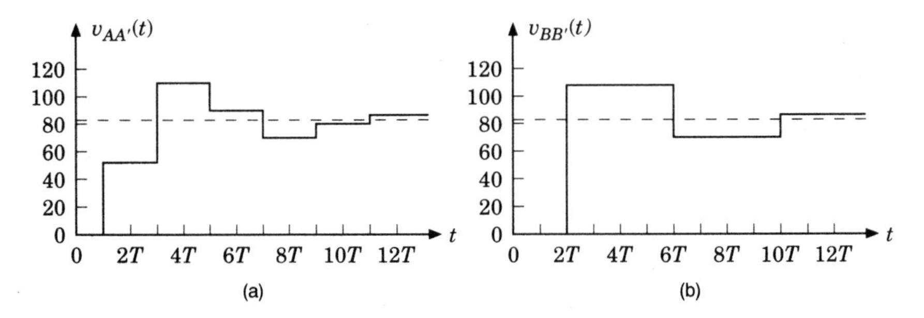

Figure 2.19 Voltage waveform at AA' (a) and at BB' (b) for the circuit of figure 2.17b.

2.8 Sources Driving Multiple Lines

As is the case for individual lines, multiple lines can also be series or parallel terminated. We see an arrangement of parallel termination in figure 2.20. It is quite obvious that the step voltage $Vu(t)$ is simultaneously impressed on all n transmission lines. The voltage propagates to the end of each line where it is absorbed without reflection. This connection does not require a great deal of study. The only drawback of this circuit is that in the high state the voltage source on the left has to supply a current equal to nV/Z_0. If the high state voltage is 5 V, as would be the case for TTL and CMOS, then for five printed-circuit-board traces with a $Z_0 = 50\,\Omega$ the line driver would have to supply a current of 500 mA. This is a very substantial (one-half ampere) current demand on the driver and also on the power supply of the system.

We now turn our attention to the series terminated arrangement shown in figure 2.21. The n transmission lines connected in parallel can be considered conceptually as one transmission line with a characteristic impedance Z_0/n. This determines the choice of the series matching resistor of value Z_0/n. The signal originating from the step voltage source $Vu(t)$ sees a resistance of value Z_0/n connected to the parallel-connected transmission-lines constituting an effective input resistance of value Z_0/n. Accordingly the initial input voltage to all the lines is $V/2$. After this voltage gets fully reflected at the right-hand open circuit, the final voltage of value V is established on all the lines.

Example 2.7 It is instructive to perform a more detailed analysis of what happens in the series terminated connection driving the two lines shown in figure 2.22. For convenience the two lines are shown to either side of the source $Vu(t)$ whose output impedance $R_S = 10\,\Omega$ has been augmented by a $15\,\Omega$ resistor to reach the value $Z_0/2$.

At time zero the step source sends out a voltage $V/2$ on both lines, as shown in the lattice diagram of figure 2.23. This voltage gets fully reflected at the open

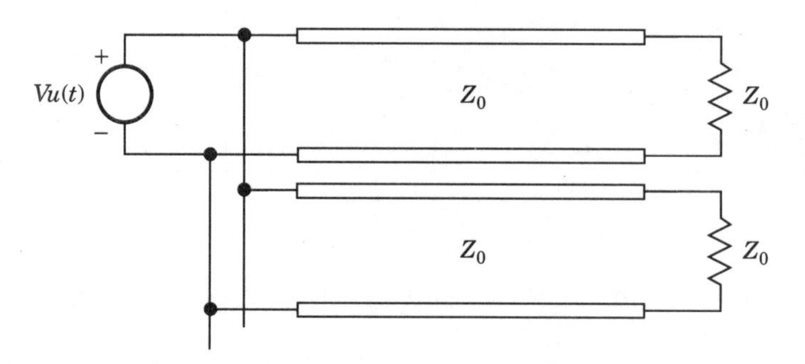

Figure 2.20 Multiple lines terminated in parallel.

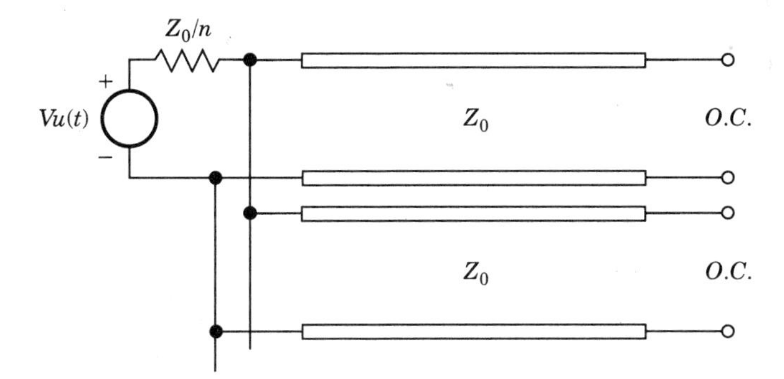

Figure 2.21 Multiple lines terminated in series.

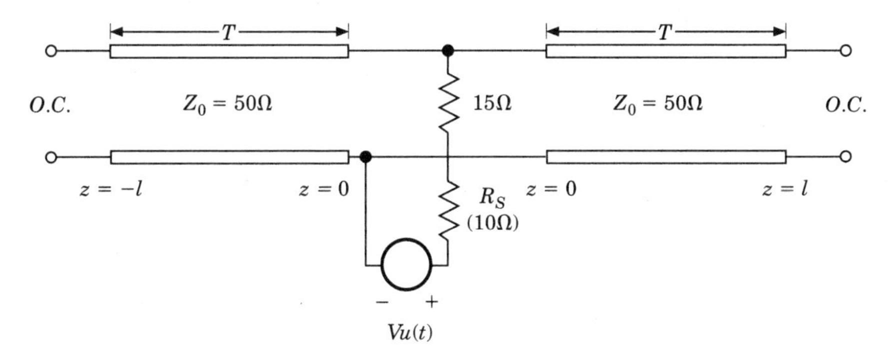

Figure 2.22 Two lines terminated in series.

circuited end of the two lines. When the reflected voltage arrives at the interface it sees an impedance of $25||50\ \Omega$. This impedance produces a reflection coefficient of $-1/2$ and a transmission coefficient of $1/2$. Accordingly the reflected waves and the transmitted waves, which occur simultaneously, cancel each other out. This action is clearly visible in the lattice diagram. ■

The above example was presented deliberately in order to study the detailed behavior of the series connection. It is left as an exercise to

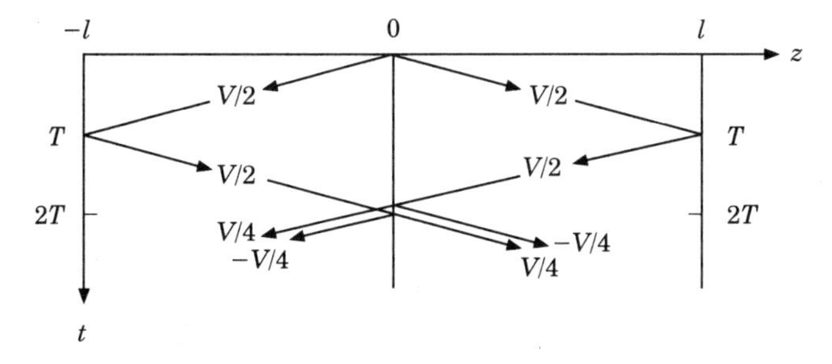

Figure 2.23 Lattice diagram for the circuit of figure 2.22.

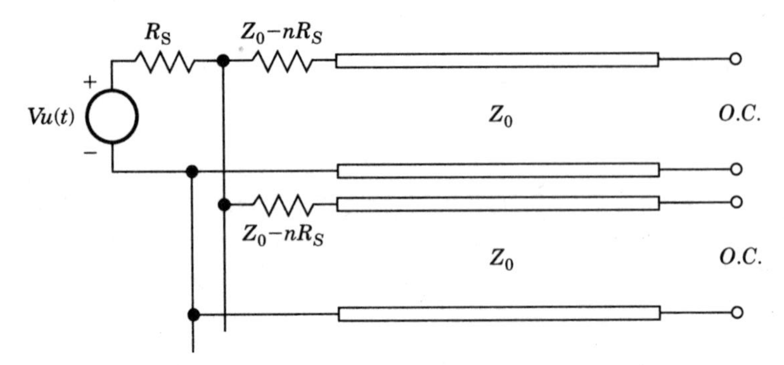

Figure 2.24 A second option for terminating multiple lines in series.

show that if the two lines are not of the same length, then the wave cancellation at the interface, which was the case in the above example, will not take place since the two reflected voltage waves will not arrive at the interface at the same time. It is quite clear that there will be some serious oscillations at the open circuited ends of the two lines. These oscillations can be reduced somewhat if, instead of using the single terminating resistor of figure 2.21, we use separate resistors for each line, as shown in figure 2.24.

To see that this is a properly matched series terminated line, we simply observe that the n resistors of value $Z_0 - nR_S$ in parallel constitute one resistor of value $Z_0/n - R_S$. This in series with the source resistor R_S makes for a total termination resistance of Z_0/n. This is a perfect series match of the n transmission lines in parallel which are equivalent to one line of characteristic impedance Z_0/n. The improvement in performance will be demonstrated in the following example.

Example 2.8 As was done in the previous example, we perform a detailed analysis of the series terminated connection driving the two lines shown in figure 2.25.

At time zero the step source sends out a voltage $V/2$ on both lines, as shown in the lattice diagram of figure 2.26. To get a better understanding of the sequence of events, we will concentrate our attention on the left line. The incident voltage gets reflected at the left open circuit and arrives at the interface AA'. Looking to the right this wave sees an impedance of $[30 + 10\|(30 + 50)] = 38.9\,\Omega$. This impedance produces a reflection coefficient of $-1/8$ and a transmission coefficient of $7/8$ at AA'. The voltage that gets reflected is therefore $-V/16$ and the voltage that is transmitted to the network to the right of AA' is $V_{AA'} = 7V/16$. In going from AA' to BB', the transmitted signal is attenuated by the factor

$$\frac{V_{BB'}}{V_{AA'}} = \frac{10\|(30 + 50)}{30 + 10\|(30 + 50)} \cdot \frac{50}{30 + 50} = \frac{1}{7}$$

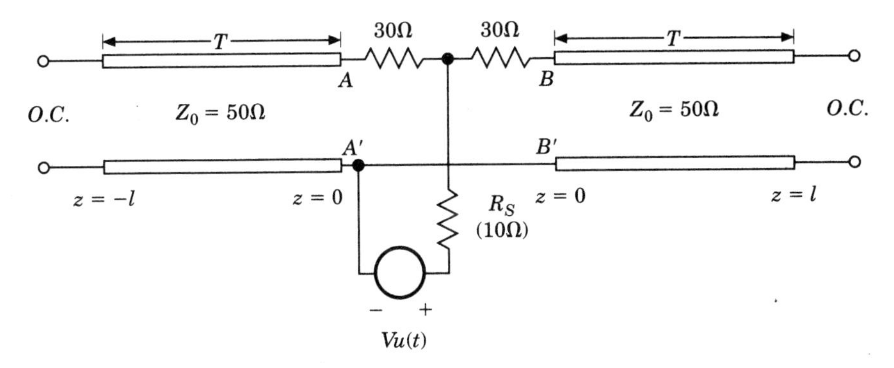

Figure 2.25 Two lines terminated in series according to the second option.

This attenuation factor multiplied by the transmitted voltage of $7V/16$ produces a voltage $V_{BB'} = V/16$. The reflected waves and the transmitted waves, which occur simultaneously, cancel each other out. But in contrast to the previous example, the two voltages being passed through the interface are 1/4 as big.

It is left as an exercise to show that if the two lines are not of the same length, then the wave cancellation at the interface, which was the case in the above example, will not take place since the two reflected voltage waves will not arrive at the interface at the same time. But since the voltages being passed through the interface are 1/4 as big in this case as they are in the previous example, the failure of the voltages to cancel is not as serious in this case. ■

2.9 One Line Driving Multiple Lines

It is of interest to investigate the behavior of a single transmission line which is at some point fanned out into a multiplicity of lines. It is very common to find this kind of geometry on printed-circuit boards of digital systems operated at relatively slow clock rates at which the transmission-line effect is not an issue. We wish to determine if this

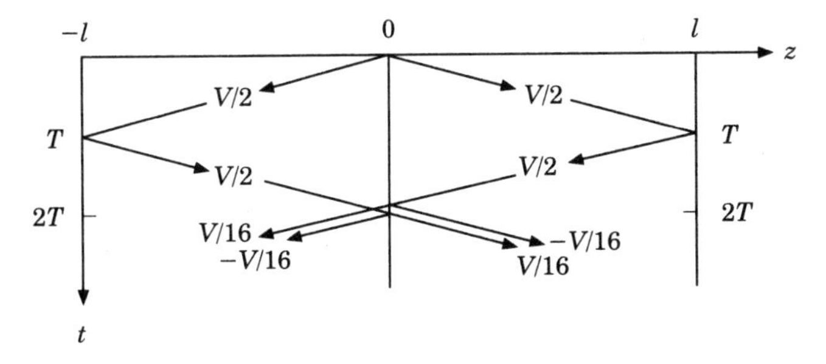

Figure 2.26 Lattice diagram for the circuit of figure 2.25.

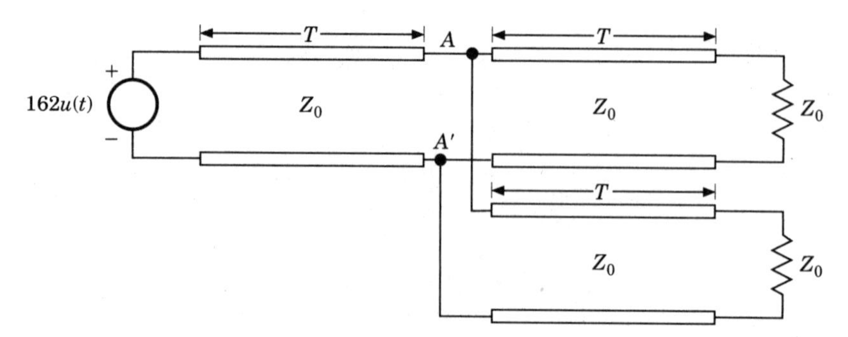

Figure 2.27 A single transmission line fanned out into two parallel terminated lines.

connection will cause problems if the clock rate is increased to the point where the printed-circuit-board traces have to be treated as transmission lines.

We begin to address this problem by examining the parallel terminated connection shown in figure 2.27. There we have a line fanned out at AA' into two other transmission lines. To facilitate the analysis all lines have the same one-way time delay T. At the interface the reflection coefficient is $-1/3$, hence the transmission coefficient is $2/3$. The 162 V wave arriving at AA' has -54 V reflected and a wave of 108 V is transmitted to both lines on the right. The lines on the right are matched, hence there are no reflections on them. But reflections do take place on the left section, as can be readily seen in the lattice diagram appearing in figure 2.28.

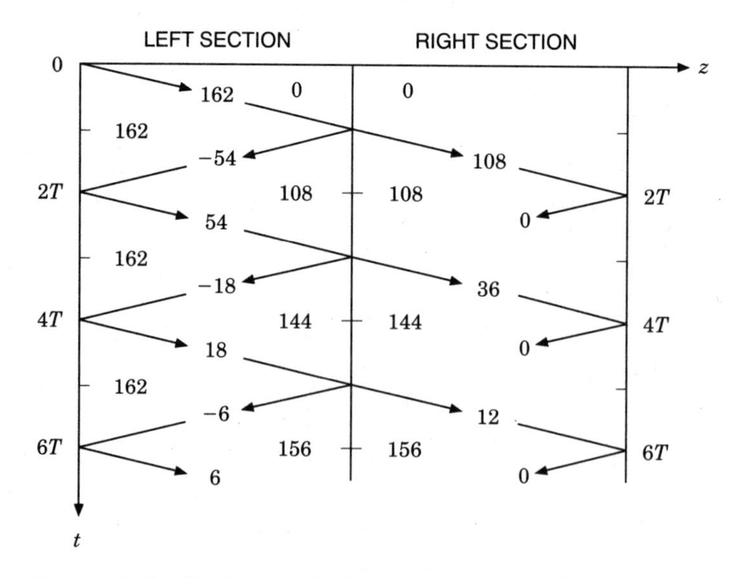

Figure 2.28 Lattice diagram for the connection shown in figure 2.27.

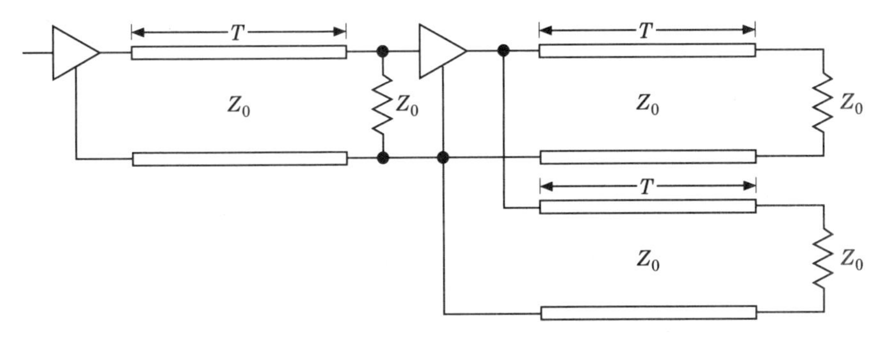

Figure 2.29 The buffered version of the circuit of figure 2.27.

We know that the steady-state voltage on all lines is 162 V. It is readily apparent from the lattice diagram that the voltage at the receiving end on the right side of figure 2.27 is a very significant 33% below the steady-state value after $2T$ seconds. One has to wait $6T$ seconds before the voltage at the loads is a modest 4% below the steady-state value. Compare this to a delay of $2T$ seconds needed for the loads to reach full steady-state in the connection shown in figure 2.29. It is clearly much better to use a buffer to isolate the two sections of transmission line.

Having looked at the shunt terminated case, we now shift our attention to the series terminated case shown in figure 2.30. As in the parallel connection, the reflection coefficient at the interface is $-1/3$; therefore the transmission coefficient is 2/3. The voltage wave launched at the source is 81 V due to the voltage division created by the series matching resistor of value Z_0. Of the 81 V arriving at the interface 2/3, or 54 V, is transmitted to the right side of the interface at AA', as shown in figure 2.31. This gets fully reflected on both of the open circuited lines. Of the 54 V returning

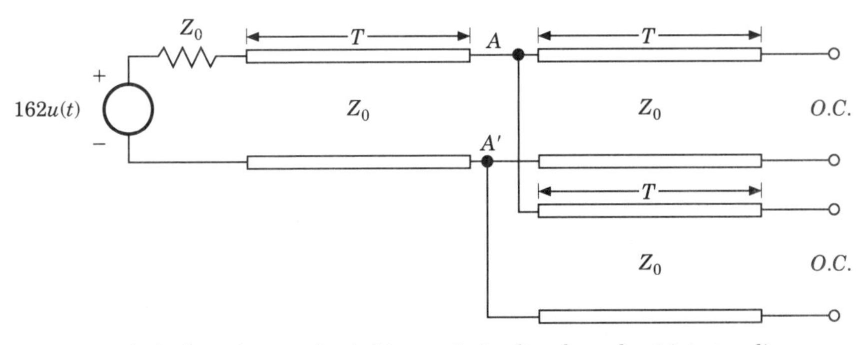

Figure 2.30 A single series terminated transmission line fanned out into two lines.

to the interface on the top line, −1/3, hence −18 V gets reflected. Of the 54 returning to the interface on the bottom line, 2/3, namely 36 V gets transmitted to the top line. Consequently the next incident wave on the top line has a value of 36 − 18 = 18 V, as seen in the lattice diagram. Since the two 54 V returning waves on the top line and the bottom line both transmit a voltage of 36 V to the left of the interface, this constitutes the 72 V wave traveling toward the left on the left section of line. The line on the left is matched, so there are no reflections on it. But reflections do take place on the right sections, as can be readily seen in the lattice diagram appearing in figure 2.31.

In the final analysis this connection behaves identically to the one shown in figure 2.27. The receiving voltage is 33% below the steady-state value after $2T$ seconds and, as in the parallel matched case, one has to wait $6T$ seconds before the voltage at the loads is 4% below the steady. The difference here is that the source supplies zero current in the high steady-state compared to the possibly large current which the source might have to supply in the high steady-state for the case depicted in figure 2.27.

The arrangement shown in figure 2.30 can be replaced by the buffered version shown in figure 2.32. Here a delay of $2T$ seconds is needed for the voltage at the loads to reach full steady-state. Each line is matched by an impedance Z_0. We have seen in the previous section that this type of match is possible when a very low output impedance buffer is used.

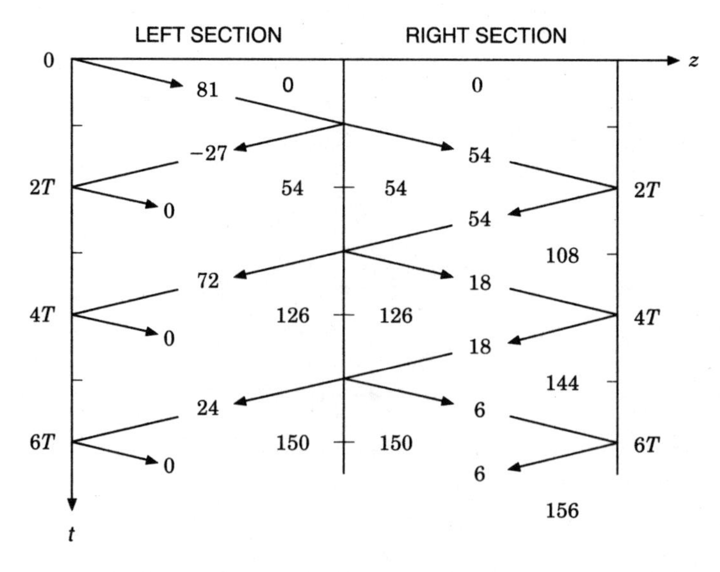

Figure 2.31 Lattice diagram for the connection shown in figure 2.30.

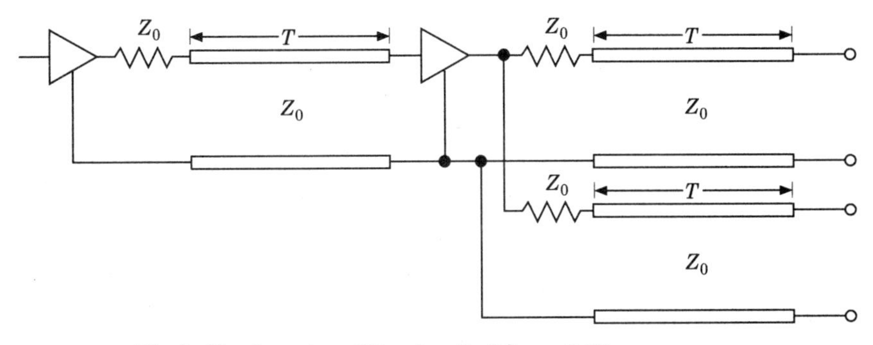

Figure 2.32 The buffered version of the circuit of figure 2.30.

2.10 Conclusion

In this chapter we put down the foundations for studying the behavior of transmission lines of finite length, namely transmission lines with terminations that cause reflections. The principal notions in that connection are the reflection coefficient ρ and the transmission coefficient τ. The idea of series and shunt terminated lines was introduced and the concept was elaborated in the sections that followed. This notion will keep recurring in later chapters. The concept of cascaded lines was elaborated in section 2.7 with further applications shown in sections 2.8 and 2.9. Our ultimate objective is to apply the theory developed to high-speed digital circuits, an undertaking that will begin with the following chapters.

Bibliography

1. W. C. Johnson, *Transmission Lines and Networks*, McGraw-Hill, New York, 1950.
2. P. C. Magnusson, *Transmission Lines and Wave Propagation*, Allyn and Bacon, Boston, Massachusetts, 1970.
3. S. R. Seshadri, *Fundamentals of Transmission Lines and Electromagnetic Fields*, Addison-Wesley, Reading, Massachusetts, 1971.

Problems

P2.1 Carry out the steps needed to derive (2.6) and (2.7).

P2.2 Using the lattice diagrams of figure 2.4, draw a voltage waveform similar to figure 2.5a

(a) For $z = l/4$.

(b) Repeat for $z = l/2$.

(c) Repeat for $z = l$.

P2.3 Using the lattice diagrams of figure 2.4, draw a current diagram similar to figure 2.5b

(a) For $z = l/4$.
(b) Repeat for $z = l/2$.
(c) Repeat for $z = l$.

P2.4 Using the lattice diagrams of figure 2.4, draw voltage and current distributions, similar to figure 2.6, for $t = 7T/2$.

P2.5 Draw the voltage and current lattice-diagrams for example 2.2. From that, draw voltage and current waveforms at $z = \frac{1}{2}l$ and $z = l$. Also draw voltage and current distributions for $t = 3T/2$.

P2.6 Draw the voltage and current lattice-diagrams for example 2.3. From that draw voltage and current waveforms at $z = \frac{1}{4}l$ and $z = l$. Also draw voltage and current distributions for $t = 3T/4$.

P2.7 Use (2.23) and (2.24) to verify the steady-state values indicated in figure 2.5. Are the results the same if (2.25) and (2.26) are used?

P2.8 Derive (2.25) and (2.26) from (2.23) and (2.24).

P2.9 In figure 2.7, we have $R_S = Z_0/2$, $R_L = 2Z_0$, and $V = 364.5$ V. Draw the lattice diagrams for this circuit and see if the voltage and current tend to the final steady-state values computed using (2.25) and (2.26).

P2.10 Consider the circuit illustrated in figure 2.23. The speed of propagation on the line is $\nu = 3 \times 10^8$ m/s and the line has a length of 3 meters. Switch SW2 is in position A, and SW1 is closed at $t = 0$. Plot the voltage distribution along the line at $t = 15$ ns and the voltage at $z = 3l/4$ as a function of time.

P2.11 Assume that in figure 2.33 steady state has been reached with SW1 closed and SW2 in position A. At $t = 0$ SW2 is thrown to position B. Plot the voltage distribution along the line at $t = 15$ ns and the voltage at $z = 3l/4$ as a function of time.

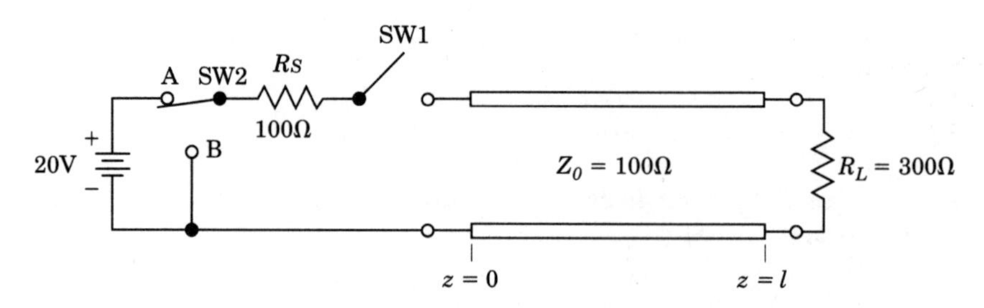

Figure 2.33 A transients problem involving switching.

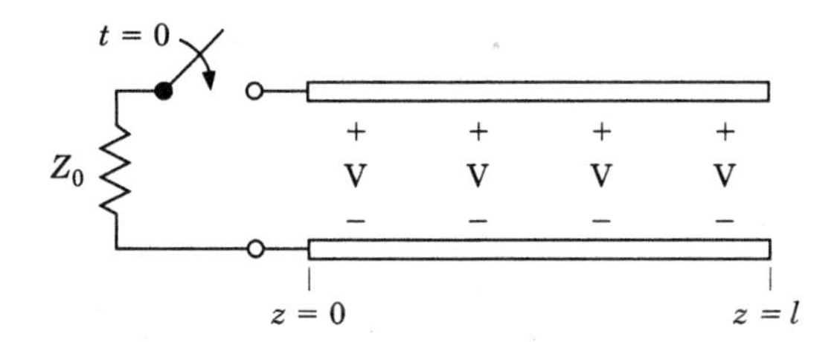

Figure 2.34 A pulse forming network.

P2.12 Assume that in figure 2.33 steady state has been reached with SW1 closed and SW2 in position A. At $t = 0$ the load resistor R_L is suddenly changed from $300\,\Omega$ to $200\,\Omega$. Determine the sequence of events after switching takes place.

P2.13 The transmission line shown in the pulse forming connection of figure 2.34 has $Z_0 = 50\,\Omega$. The line is first charged to V volts. The switch is closed at $t = 0$. Analyze the resultant transient problem using a lattice diagram. The line has a length l and the speed of propagation is ν. How can the duration of the pulse be doubled?

P2.14 Consider the series and parallel terminations of figure 2.13 combined so that a line is terminated at both ends in its characteristic impedance Z_0. Assume that the left gate sends a step voltage $Vu(t)$. Give arguments for and against this method of termination.

P2.15 In figure 2.13a, assume that the left gate sends a step voltage $Vu(t)$ and that $R_S = Z_0$. Draw a lattice diagram and plot the waveforms $v(\frac{1}{4}l, t)$, $v(\frac{1}{2}l, t)$, and $v(l, t)$.

P2.16 In figure 2.13b, assume that the left gate sends a step voltage $Vu(t)$ and that $R_L = Z_0$. Draw a lattice diagram and plot the waveforms $v(\frac{1}{4}l, t)$, $v(\frac{1}{2}l, t)$, and $v(l, t)$.

P2.17 Complete labeling of the diagonal traces in figure 2.16 and indicate the voltages on both sides of the interface after the arrival of each wave.

P2.18 In example 2.6 the positive 81 V step is followed after $2T$ seconds by a negative 81 V step. This is equivalent to launching an 81 V pulse of duration $2T$.

(a) Use superposition to determine the voltage waveform at AA' for the circuit of figure 2.17b. Are there large negative undershoots at the input to the gate? Is the positive level of 81 V ever reached at the gate inputs connected at AA'?

(b) Repeat the above analysis for a pulse duration of $4T$ seconds.

P2.19 Assume in example 2.7 that the transmission line on the right is 1.5 times as long as the transmission line on the left. Draw a lattice diagram of the circuit and sketch the voltage at the open-circuit loads of both lines.

P2.20 Assume in example 2.8 that the transmission line on the right is 1.5 times as long as the transmission line on the left. Draw a lattice diagram of the circuit and sketch the voltage at the open-circuit loads of both lines. Is the voltage better behaved in this case than it was in problem 2.19?

P2.21 Assume in example 2.8 that the voltage source has zero output resistance, namely $R_S = 0$. Assume also that the transmission line on the right is 1.5 times as long as the transmission line on the left. Using the system of figure 2.24 of terminating lines, draw a lattice diagram of the circuit and sketch the voltage at the open-circuit loads of both lines. Is the voltage better behaved in this case than it would be if a single terminating resistor were placed as in figure 2.21?

P2.22 Instead of the two lines shown, we have n lines connected in parallel to the right of AA' in figure 2.27. Write the expression for reflection coefficient at AA'. How does it behave as n becomes very large? Does increasing n improve on the behavior of this connection?

Chapter

3

Transients on Transmission Lines

3.1 Introduction

The signals that are transferred between integrated circuits and between subassemblies in digital systems can be considered to be pulses with either abrupt (step) transitions, or pulses with short rise times. Some of the problems that we are interested in solving pertain to the placement of integrated circuits along the connecting printed-circuit board paths, which constitute transmission lines. Each integrated circuit has a small amount of parasitic capacitance at the input to each gate, which will cause a slowing down of the pulse transitions. To examine the behavior of the transmission line in the presence of capacitance at various locations, there will be a need to employ Laplace transforms. The reader is referred to appendix B for a review of the subject if that is needed.

In this chapter we wish to demonstrate the use of the Laplace transform in determining the transient behavior of both series and parallel terminated lines in the presence of capacitive loading. As mentioned above, the capacitive loads represent the parasitic capacitance that is present at the input of each digital gate. Another objective is to determine if it is better to place all the receiving-gate inputs at the end of a transmission line, or to distribute them in a uniform manner along the length of the line. The criterion for determining the benefit of one method over another will be the resultant speed of the voltage transitions at the inputs of the gates. In figure 3.1

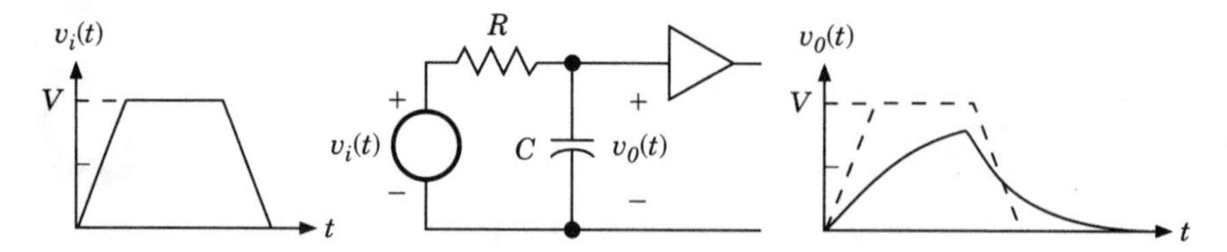

Figure 3.1 Illustration of rise-time degradation due to an R–C circuit.

we have a trapezoidal pulse $v_i(t)$ applied to an R–C circuit whose output voltage $v_o(t)$ is the input voltage to a digital gate. The R–C circuit causes rise-time degradation in the trapezoidal pulse. If the time constant of the R–C circuit is great enough compared to the duration of the pulse, then we can see from the illustration that the rise-time degradation can become so severe that the voltage $v_o(t)$ may not reach a sufficient value to cause the gate to recognize a transition from a low to a high state. Hence rise-time degradation can have a very deleterious effect on the operation of digital circuits. By contrast, a distortionless time delay in the pulse, such as the kind that is caused by a lossless transmission line, will slow the operation of the circuit, but at least there will be no errors in the interpretation of the binary data. It will be seen later that the circuit of 3.1 represents fairly closely the circuits that will be discussed in this chapter.

Before we proceed we will define two concepts. *Time domain reflection* (TDR) describes the observation of waveforms at the input end of a transmission system for the purpose of gaining information pertaining to the configuration of the system. *Time domain transmission* (TDT) describes the observation of waveforms at the load end of a transmission system for the same purpose. Both TDR and TDT will be discussed, although it may turn out that for some systems only one of the two observations is feasible due to some physical system constraint.

The concepts of series and parallel termination of transmission lines were covered in chapter 2. These methods of termination are complicated by the fact that each gate input that is connected to the transmission line has a small amount of parasitic capacitance across its input terminals. We will now analyze the effect that this has on the TDR and TDT signals for the two methods of termination.

3.2 Capacitive Load — Series Match

We will start by considering the transmission line configuration shown in figure 3.2. The circuit contains a lossless transmission-line terminated at the source in a matched impedance Z_0 and at the load with

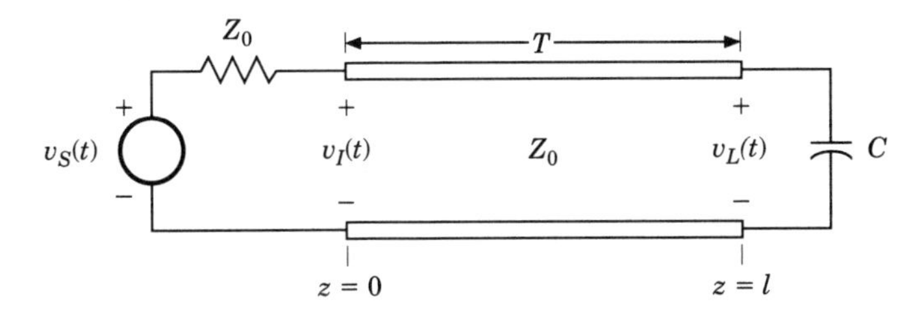

Figure 3.2 Series terminated line with capacitive load.

an open circuit in parallel with a capacitance C. The load impedance is a function of the Laplace transform variable s, and is given by

$$Z_L(s) = \frac{1}{sC} \tag{3.1}$$

The reflection coefficient of (2.8) has to be modified for our purposes by replacing R_L by $Z_L(s)$ as below

$$\rho_L(s) = \frac{Z_L(s) - Z_0}{Z_L(s) + Z_0} \tag{3.2}$$

The above modification generalizes (2.8) to apply to loads that are reactive, rather than just being resistive. We find upon substitution of (3.1) into (3.2) that

$$\rho_L(s) = -\frac{s - 1/t_s}{s + 1/t_s} \tag{3.3}$$

where the time constant t_s, for the series terminated circuit, is defined by

$$t_s \equiv CZ_0 \tag{3.4}$$

It is useful to find the transmission coefficient

$$\tau_L(s) = 1 + \rho_L(s) \tag{3.5}$$

because it determines how much of the incident voltage is transferred to the load. Using (3.3) in (3.5), we obtain

$$\tau_L(s) = \frac{2/t_s}{(s + 1/t_s)} \tag{3.6}$$

With these preliminaries out of the way we now proceed to analyze the problem of figure 3.2. Figure 3.3 shows in a schematic manner how

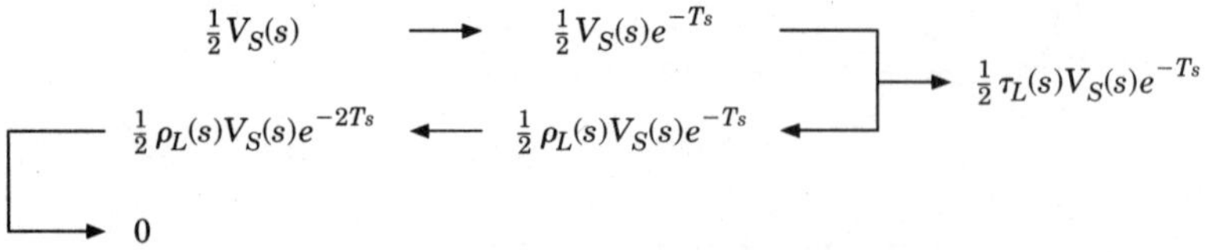

$\frac{1}{2}V_S(s)$ → $\frac{1}{2}V_S(s)e^{-Ts}$ → $\frac{1}{2}\tau_L(s)V_S(s)e^{-Ts}$

$\frac{1}{2}\rho_L(s)V_S(s)e^{-2Ts}$ ← $\frac{1}{2}\rho_L(s)V_S(s)e^{-Ts}$

→ 0

Figure 3.3 A lattice-like diagram for the circuit of figure 3.2.

the signal, which is launched on the left side of the line, goes through reflections as it encounters the terminations at both ends of the line.

To arrive at figure 3.3 we observe that a signal $V_S(s)$ which is introduced on the left side of the line produces a first incident voltage of one half that value. This is due to the voltage division which takes place between the source resistance Z_0 and the input impedance to the line which is also Z_0. This first wave arrives at the load after a delay of T seconds. The exponential factor e^{-Ts} multiplying $V_S(s)$ accounts for this delay. Multiplying this incident voltage by $\tau_L(s)$ yields the load voltage which is shown on the right side of the diagram. Multiplying the incident voltage by $\rho_L(s)$ yields the reflected voltage which is shown heading back toward the sending end. The reflected voltage finds a matched load (with an associated reflection coefficient of zero) at the sending end, so there are no further reflections to consider.

From figure 3.3 we can readily write the voltage at the load as

$$V_L(s) = \frac{1}{2}\tau_L(s)V_S(s)e^{-Ts} \tag{3.7}$$

From figure 3.3 we can also write the voltage at the input to the line as

$$V_I(s) = \frac{1}{2}V_S(s) + \frac{1}{2}\rho_L(s)V_S(s)e^{-2Ts} \tag{3.8}$$

Circuit of Figure 3.2 — Response to a Step Voltage

We will first proceed to solve the above equations for a step voltage input to the line, namely

$$v_S(t) = u(t) \tag{3.9}$$

The choice of a step input voltage is made to facilitate the analysis and to give us some immediate insight into the transient problem. The Laplace transform of (3.9) is

$$V_S(s) = \frac{1}{s} \tag{3.10}$$

Substituting the above, and (3.3) and (3.6) into (3.7) and (3.8) we get the expressions

$$V_L(s) = \frac{1/t_s}{s(s + 1/t_s)} e^{-Ts} \tag{3.11}$$

$$V_I(s) = \frac{1}{2s} - \frac{s - 1/t_s}{2s(s + 1/t_s)} e^{-2Ts} \tag{3.12}$$

When the above two expressions are expanded into partial-fraction form we have

$$V_L(s) = \left[\frac{1}{s} - \frac{1}{s + 1/t_s}\right] e^{-Ts} \tag{3.13}$$

$$V_I(s) = \frac{1}{2s} + \left[\frac{1}{2s} - \frac{1}{s + 1/t_s}\right] e^{-2Ts} \tag{3.14}$$

From the last two equations we readily obtain the time domain expressions

$$v_L(t) = \left[1 - e^{-(t-T)/t_s}\right] u(t - T) \tag{3.15}$$

$$v_I(t) = \frac{1}{2} u(t) + \left[\frac{1}{2} - e^{-(t-2T)/t_s}\right] u(t - 2T) \tag{3.16}$$

The above results are the TDT and TDR waveforms, respectively, for the system shown in figure 3.2. The results of the above two equations are plotted in figure 3.4.

The step function waveform that was launched on the left side of the circuit in figure 3.2 had a rise time of zero. In an ideal system, without the capacitance C, the step function would appear at the load T seconds later. With the capacitance C present we have an exponential waveform at the output, beginning at time T. Clearly we have experienced a *rise-time degradation* in the transmitted signal. This is measured as the *increase in the 50% point delay* due to capacitive loading, indicated as t_a in figure 3.4a. This time delay can be found by setting $v_L(t)$ in (3.15) equal to 0.5. The result in this case is

$$t_a \equiv \text{increase in the 50\% point delay} = 0.693 t_s \tag{3.17}$$

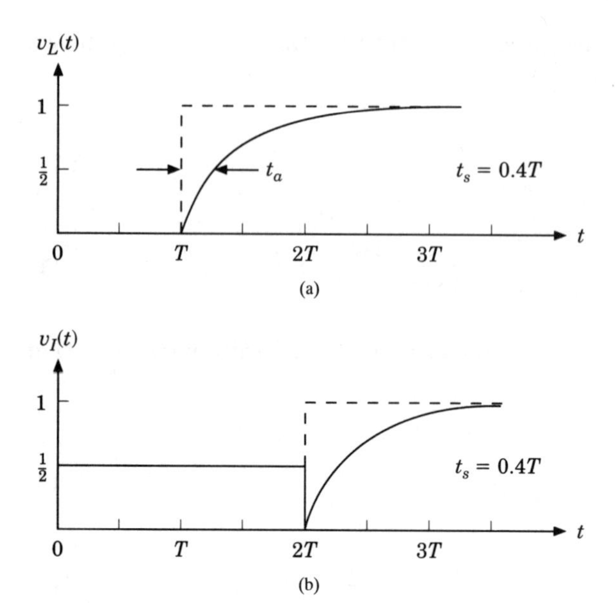

Figure 3.4 (a) TDT waveform and (b) TDR waveform for the circuit of figure 3.2.

Circuit of Figure 3.2 — Response to a Step Voltage with Ramp Transition

A step function excitation was chosen for simplicity in the analysis that was just concluded. It simplified the solution substantially and gave us the opportunity to gain some insight into the behavior of a capacitively loaded transmission-line. The voltage source that is more representative of waveforms in real digital systems is the unit step function with ramp transition defined in (3.18). It accounts for the fact that all digital waveforms require non-zero time intervals to make their voltage transitions.

$$u_r^a(t) \equiv \begin{cases} 0, & \text{for } t < 0 \\ t/a, & \text{for } 0 \leq t < a \\ 1, & \text{for } a \leq t \end{cases} \tag{3.18}$$

The parameter a represents the *ramp transition time*. It needs to be made clear that this parameter must be distinguished from the signal *rise time* t_r, which is the time required for the signal to go from 10% to 90% of its final value. All of the above parameters are illustrated in figure 3.5.

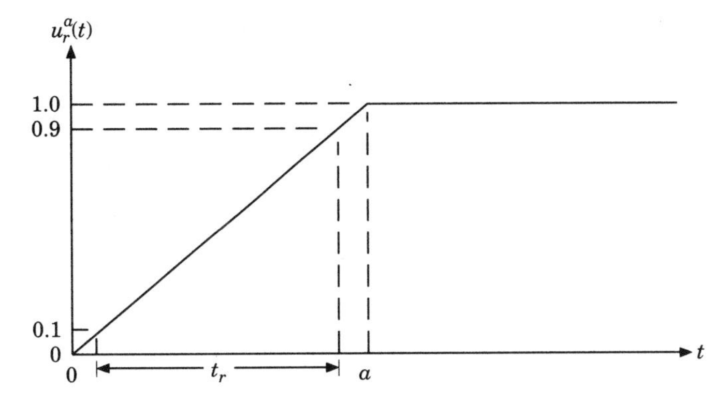

Figure 3.5 Unit step function with ramp transition.

The unit step function with ramp transition can be written as the difference of two unit ramp functions as

$$u_r^a(t) = \frac{t}{a}u(t) - \frac{t - a}{a}u(t - a) \tag{3.19}$$

The Laplace transform of the above can be found by consulting the tables in appendix B, with the result

$$U_r^a(s) = \frac{1}{as^2}(1 - e^{-as}) \tag{3.20}$$

We will now repeat the analysis of the circuit in figure 3.2 using

$$v_S(t) = u_r^a(t) \tag{3.21}$$

so that

$$V_S(s) = \frac{1}{as^2}(1 - e^{-as}) \tag{3.22}$$

Substituting the above, and (3.3) and (3.6) into (3.7) and (3.8) we get the expressions

$$V_L(s) = \frac{1}{at_s s^2(s + 1/t_s)}(1 - e^{-as})e^{-Ts} \tag{3.23}$$

$$V_I(s) = \frac{1}{2as^2}(1 - e^{-as}) - \frac{1}{2as^2}\frac{s - 1/t_s}{s + 1/t_s}(1 - e^{-as})e^{-2Ts} \tag{3.24}$$

When the above two expressions are expanded into partial fraction expression form we have

$$V_L(s) = \left[\frac{1}{as^2} - \frac{t_s}{as} + \frac{t_s}{a(s + 1/t_s)}\right](1 - e^{-as})e^{-Ts} \tag{3.25}$$

$$\begin{aligned} V_I(s) = {} & \frac{1}{2as^2}(1 - e^{-as}) \\ & + \left[\frac{1}{2as^2} - \frac{t_s}{as} + \frac{t_s}{a(s + 1/t_s)}\right](1 - e^{-as})e^{-2Ts} \end{aligned} \tag{3.26}$$

From the last two equations we readily obtain the time domain expressions

$$\begin{aligned} v_L(t) = {} & \left\{\frac{t - T}{a} - \frac{t_s}{a}\left[1 - e^{-(t-T)/t_s}\right]\right\}u(t - T) \\ & - \left\{\frac{t - T - a}{a} - \frac{t_s}{a}\left[1 - e^{-(t-T-a)/t_s}\right]\right\}u(t - T - a) \end{aligned} \tag{3.27}$$

$$\begin{aligned} v_I(t) = {} & \frac{t}{2a}u(t) - \frac{t - a}{2a}u(t - a) \\ & + \left\{\frac{t - 2T}{2a} - \frac{t_s}{a}\left[1 - e^{-(t-2T)/t_s}\right]\right\}u(t - 2T) \\ & - \left\{\frac{t - 2T - a}{2a} - \frac{t_s}{a}\left[1 - e^{-(t-2T-a)/t_s}\right]\right\}u(t - 2T - a) \end{aligned} \tag{3.28}$$

We can now use (3.27) and (3.28) to plot the TDT and the TDR waveforms for the response of the circuit shown in figure 3.2 to a unit step function with ramp transition. The curves appear in figure 3.6. The step-function waveform with ramp transition that was launched on the left side of the circuit in figure 3.2 would produce the waveforms shown by the dashed lines if the circuit had no capacitance at the output. The solid lines show the waveforms with the capacitance present. The TDR waveform at the input to the transmission line is shown in figure 3.6b. Figure 3.6a shows the TDT waveform that occurs at the output of the transmission line. It is clear that the signal has suffered a rise-time degradation.

The increase in the 50% point delay due to capacitive loading is indicated as t_a in figure 3.6a. This time delay can be found by calculating the time at which $v_L(t)$ in (3.27) equals 0.5 and then subtracting half the ramp time. The result is the graph of normalized time delay t_a/t_r

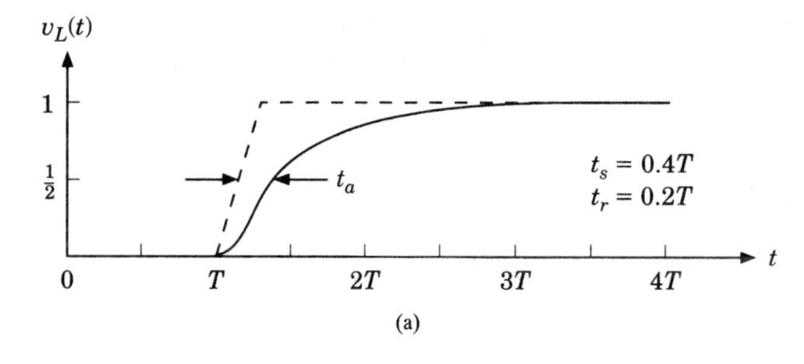

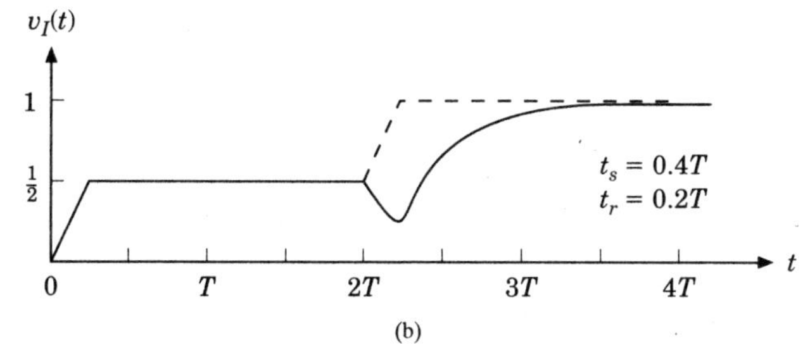

Figure 3.6 TDT waveform (a) and TDR waveform (b).

versus normalized circuit time constant t_s/t_r appearing in figure 3.7. In practice slightly greater values of t_a may be observed due to the fact that the waveforms transmitted by digital logic devices usually do not have the ideal shape shown in figure 3.5.

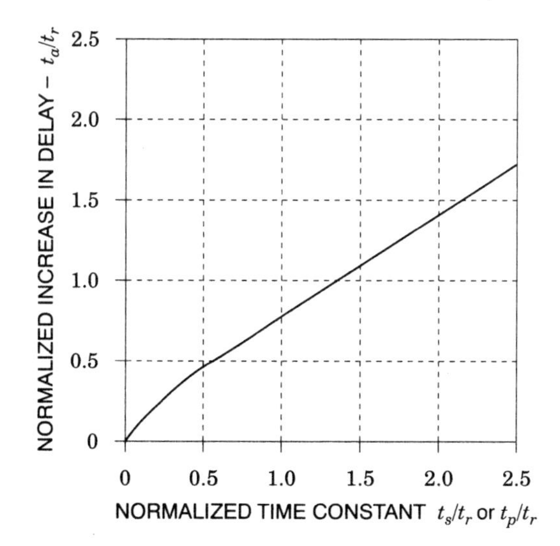

Figure 3.7 Normalized 50% point delay t_a/t_r versus normalized circuit time constant t_s/t_r or t_p/t_r.

3.3 Capacitive Load — Parallel Match

In this analysis we will essentially retrace the steps of section 3.2 for the parallel terminated line shown in figure 3.8. The circuit contains a lossless transmission line terminated with a load consisting of a matching impedance Z_0 in parallel with a capacitance C. Since the circuit is driven by an ideal voltage source, then we know from the start that the voltage at the input to the line remains $v_S(t)$ irrespective of reflections. Accordingly, this circuit possesses no TDR waveform, and our work will be cut in half.

The transmission-line load is a function of the Laplace transform variable s, and is given by

$$Z_L(s) = Z_0 \left\| \frac{1}{sC} = \frac{Z_0}{sCZ_0 + 1} \right. \tag{3.29}$$

Using the above in (3.2) we find

$$\rho_L(s) = \frac{-s}{s + 1/t_p} \tag{3.30}$$

where the time constant t_p, of the parallel terminated circuit, is defined by

$$t_p \equiv \frac{CZ_0}{2} \tag{3.31}$$

We observe that for the same values of circuit parameters the time constant in (3.31) is one half the value of the one in (3.4). So events should happen faster in the parallel terminated circuit as compared to the series terminated circuit.

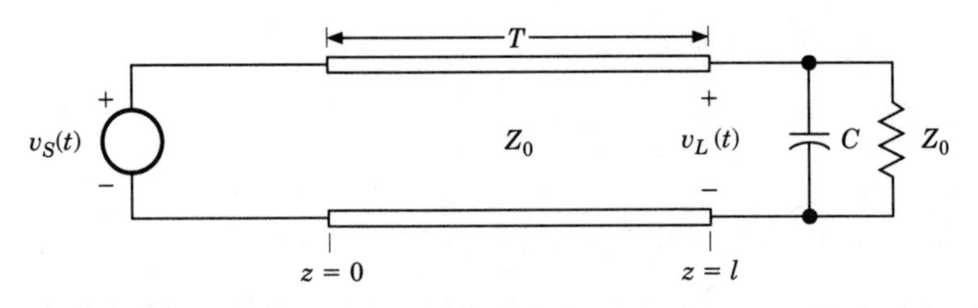

Figure 3.8 Parallel terminated line with capacitive load.

Substituting (3.30) into (3.5) we find the transmission coefficient

$$\tau_L(s) = \frac{1/t_p}{s + 1/t_p} \tag{3.32}$$

Figure 3.9 shows in a schematic manner how the signal, which is launched on the left side of the line, goes through reflections as it encounters the terminations at both ends of the line. Since there is an ideal source connected at the sending end, the reflection coefficient there is -1. This is in contrast to the series terminated case, where the transmission line was matched at the sending end. We can therefore expect many reflections for the parallel terminated case. We now proceed to analyze the problem of figure 3.8.

From figure 3.9 we can readily write the voltage at the load as the sum

$$V_L(s) = \tau_L(s)\Big[e^{-Ts} - \rho_L(s)e^{-3Ts} + \rho_L^2(s)e^{-5Ts} - \cdots\Big]V_S(s) \tag{3.33}$$

Circuit of Figure 3.8 — Response to a Step Voltage

We will first analyze the case of a step function source driving the transmission line of figure 3.8. We substitute its Laplace transform found in (3.10), as well as (3.30) and (3.32) into (3.33), to obtain

$$V_L(s) = \frac{1/t_p}{s + 1/t_p} \times \left[e^{-Ts} + \frac{s}{s + 1/t_p}e^{-3Ts} + \left(\frac{s}{s + 1/t_p}\right)^2 e^{-5Ts} + \cdots\right]\frac{1}{s} \tag{3.34}$$

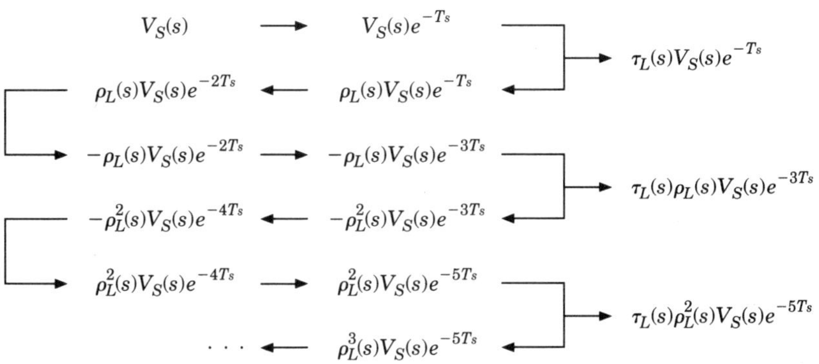

$V_S(s) \longrightarrow V_S(s)e^{-Ts}$ $\longrightarrow \tau_L(s)V_S(s)e^{-Ts}$

$\rho_L(s)V_S(s)e^{-2Ts} \longleftarrow \rho_L(s)V_S(s)e^{-Ts}$

$-\rho_L(s)V_S(s)e^{-2Ts} \longrightarrow -\rho_L(s)V_S(s)e^{-3Ts}$ $\longrightarrow \tau_L(s)\rho_L(s)V_S(s)e^{-3Ts}$

$-\rho_L^2(s)V_S(s)e^{-4Ts} \longleftarrow -\rho_L^2(s)V_S(s)e^{-3Ts}$

$\rho_L^2(s)V_S(s)e^{-4Ts} \longrightarrow \rho_L^2(s)V_S(s)e^{-5Ts}$ $\longrightarrow \tau_L(s)\rho_L^2(s)V_S(s)e^{-5Ts}$

$\cdots \longleftarrow \rho_L^3(s)V_S(s)e^{-5Ts}$

Figure 3.9 A lattice-like diagram for the circuit of figure 3.8.

After we multiply out the above expression, we find that all terms on the right side of the above equation are ready for inversion into the time domain with the exception of the first term on the right. Expanding that term into a partial fraction expansion, we obtain

$$V_L(s) = \left[\frac{1}{s} - \frac{1}{s + 1/t_p}\right]e^{-Ts} + \frac{1/t_p}{(s + 1/t_p)^2}e^{-3Ts} + \frac{(1/t_p)s}{(s + 1/t_p)^3}e^{-5Ts} + \cdots \quad (3.35)$$

The time domain equivalents of the first two terms on the right are found quite readily by consulting table B.1 in appendix B. The inversion of the last term on the right requires a little elaboration. Again using table B.1 we readily find

$$\mathcal{L}\left[\frac{1}{2t_p}(t - 5T)^2 e^{-(t-5T)/t_p}u(t - 5T)\right] = \frac{1/t_p}{(s + 1/t_p)^3}e^{-5Ts} \quad (3.36)$$

The time differentiation theorem taken from table B.2 in appendix B is restated below for convenience.

$$\mathcal{L}\left[\frac{dx(t)}{dt}\right] + x(0-) = sX(s) \quad (3.37)$$

The time function appearing in (3.36) evaluates to zero at $t = 0-$. Therefore when (3.37) is applied to (3.36), it is only necessary to differentiate the time function on the left side to obtain

$$\mathcal{L}\left\{\left[\frac{t - 5T}{t_p} - \frac{1}{2}\left(\frac{t - 5T}{t_p}\right)^2\right]e^{-(t-5T)/t_p}u(t - 5T)\right\} = \frac{(1/t_p)s}{(s + 1/t_p)^3}e^{-5Ts} \quad (3.38)$$

Now that we have the above result, we can readily write the inverse Laplace transform of (3.35)

$$v_L(t) = \left[1 - e^{-(t-T)/t_p}\right]u(t - T) + \frac{t - 3T}{t_p}e^{-(t-3T)/t_p}u(t - 3T) + \left[\frac{t - 5T}{t_p} - \frac{1}{2}\left(\frac{t - 5T}{t_p}\right)^2\right]e^{-(t-5T)/t_p}u(t - 5T) + \cdots \quad (3.39)$$

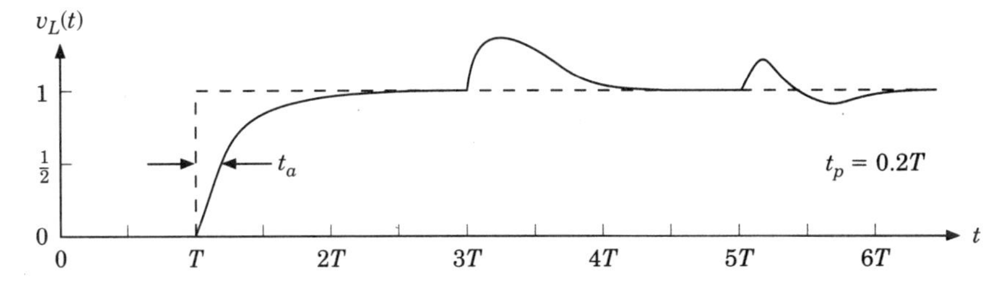

Figure 3.10 TDT waveform for the circuit of figure 3.8.

The above result is the TDT waveform for the circuit shown in figure 3.8. It is plotted in figure 3.10. In this case the source resistance is zero, hence the reflection coefficient at the source is -1. As a consequence we get additional reflections arriving at the output, each causing a minor bump in the TDT waveform. Most digital logic devices will not be damaged by the voltage overshoot of 37%.

The rise-time degradation, indicated by t_a in figure 3.10, is found by setting $v_L(t)$ of (3.39) equal to 0.5. The result, normalized with respect to t_p, is the same in this case as it was for the series terminated circuit that is given in (3.17). *It must be noted, however, that for the same circuit values, the time constant t_p for the parallel damped circuit is one half that of the time constant of a series damped circuit, so the rise-time degradation in absolute terms is smaller by one half in this case.*

Circuit of Figure 3.8 — Response to a Step Voltage with Ramp Transition

Having successfully concluded the analysis of figure 3.8 for a step function excitation we are now ready to examine its behavior when excited by the unit step function with ramp transition defined in (3.18). Substituting (3.22), as well as (3.30) and (3.32) into (3.33), we obtain

$$V_L(s) = \frac{1/t_p}{s + 1/t_p}\left[e^{-Ts} + \frac{s}{s + 1/t_p}e^{-3Ts} + \left(\frac{s}{s + 1/t_p}\right)^2 e^{-5Ts} + \cdots\right] \times \frac{1}{as^2}(1 - e^{-as}) \tag{3.40}$$

The above is multiplied out to obtain the form below, which is suitable for partial fraction expansion

$$V_L(s) = \left[\frac{1}{at_p s^2(s + 1/t_p)}e^{-Ts} + \frac{1}{at_p s(s + 1/t_p)^2}e^{-3Ts} + \frac{1}{at_p(s + 1/t_p)^3}e^{-5Ts} + \cdots\right](1 - e^{-as}) \tag{3.41}$$

In the above equation, the third term on the right is in partial fraction form. The first two terms, however, need to be expanded into partial fraction form. The resultant equation, ready for inversion, is

$$V_L(s) = \Bigg\{\left[\frac{1/a}{s^2} - \frac{t_p/a}{s} + \frac{t_p/a}{s + 1/t_p}\right]e^{-Ts} - \left[\frac{1/a}{(s + 1/t_p)^2} - \frac{t_p/a}{s} + \frac{t_p/a}{s + 1/t_p}\right]e^{-3Ts} + \frac{1}{at_p(s + 1/t_p)^3}e^{-5Ts} + \cdots\Bigg\}(1 - e^{-as}) \tag{3.42}$$

From the last equation we readily obtain the time domain expression

$$\begin{aligned} v_L(t) = {} & \left\{\frac{t - T}{a} - \frac{t_p}{a}\left[1 - e^{-(t-T)/t_p}\right]\right\}u(t - T) \\ & - \left\{\frac{t - T - a}{a} - \frac{t_p}{a}\left[1 - e^{-(t-T-a)/t_p}\right]\right\}u(t - T - a) \\ & - \left\{\frac{t - 3T}{a}e^{-(t-3T)/t_p} - \frac{t_p}{a}\left[1 - e^{-(t-3T)/t_p}\right]\right\}u(t - 3T) \\ & + \Bigg\{\frac{t - 3T - a}{a}e^{-(t-3T-a)/t_p} \\ & \qquad \frac{t_p}{a}\left[1 - e^{-(t-3T-a)/t_p}\right]\Bigg\}u(t - 3T - a) \\ & + \frac{(t - 5T)^2}{2at_p}e^{-(t-5T)/t_p}u(t - 5T) \\ & - \frac{(t - 5T - a)^2}{2at_p}e^{-(t-5T-a)/t_p}u(t - 5T - a) + \cdots \end{aligned} \tag{3.43}$$

We can now use (3.43) to plot the TDT waveform for the response of the circuit shown in figure 3.8 to a unit step function with ramp transition. The curve appears in figure 3.11.

The step function waveform with ramp transition that was launched on the left side of the circuit in figure 3.8 would produce the waveform

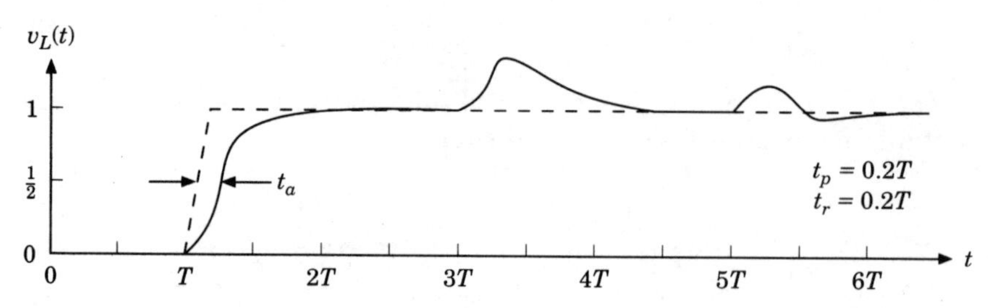

Figure 3.11 TDT waveform for the circuit of figure 3.8, for a step function with ramp transition.

shown by the dashed lines if the circuit had no capacitance at the output. The solid lines show the waveform with the capacitance present. It is clear that the signal has suffered a rise-time degradation.

The increase in the 50% point delay due to capacitive loading is indicated as t_a in figure 3.11. Comparing the terms multiplying $u(t - T)$ and $u(t - T - a)$ in (3.43) and (3.27) we see that they are identical. Hence t_a for the series and parallel terminated circuits is identical for the same time constants t_s and t_p. Therefore figure 3.7 can be used to find the excess time delay for both series and shunt terminated circuits.

It was previously pointed out that the time constant t_p for the parallel terminated circuit is one half that of the time constant of the series terminated circuit. This is because in the parallel terminated circuit the capacitance C is shunted by a resistance of value Z_0 on each side. In the series terminated circuit the capacitance C is only shunted by a resistance Z_0. So all other parameters being the same, the parallel terminated circuit will have approximately one half the rise-time degradation that one gets in the series terminated circuit. But we will see in chapter 6 that there are other reasons for preferring one method of termination over the other.

3.4 Distributed Capacitive Load — Series Match

As was mentioned in the introduction to this chapter, each digital-logic gate-input that is connected to a printed-circuit-board path possesses a small amount of input capacitance. The question that we want to address is whether it is better to lump all gate-inputs at the load end of a transmission line, or to distribute them in a uniform manner along the length of the line. To find an answer to this question we will study the step response of the three circuits shown in figure 3.12. Figure 3.12a shows a series-terminated transmission line with a capacitor connected at the load. We can think of this capacitor as representing the total input-capacitance of a number of gates connected to the end of this line. In figure 3.12b the capacitor has been split into two parts and distributed between equal lengths of the transmission line, as would be done if one half of the gates were moved half way back toward the signaling gate. The process has been carried a step further in figure 3.12c where the capacitor has been divided into three parts and distributed uniformly along the line.

The portion of the solution of interest is the one that will show the rise-time degradation. We have seen in the previous section that the first incident wave arriving at the load is the one that allows us to determine the rise-time degradation. We will therefore restrict our

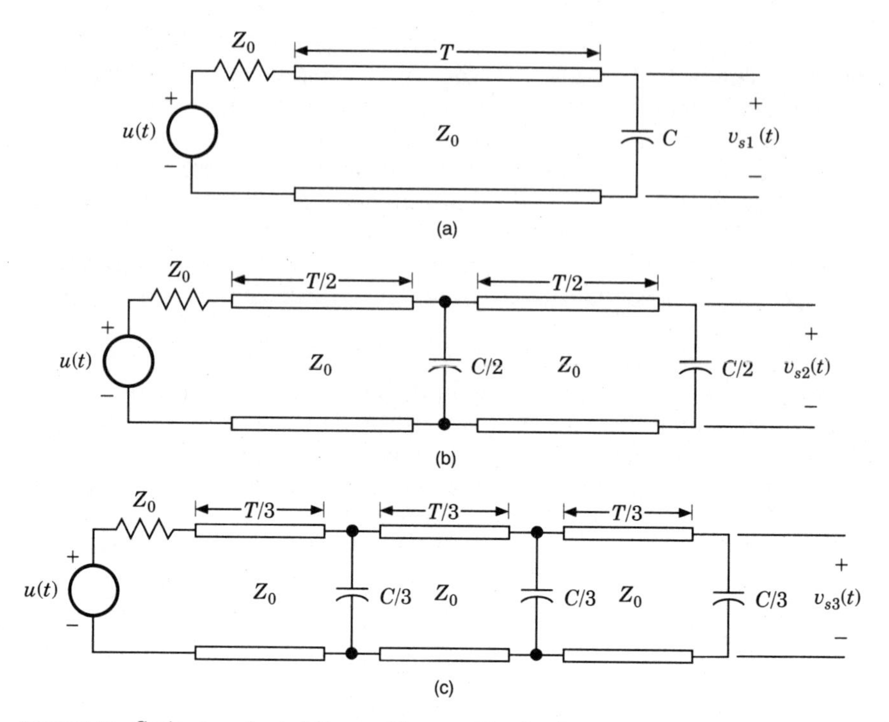

Figure 3.12 Series terminated lines with capacitive loads.

attention to the solution applicable to the arrival of the first incident wave.

In each case depicted in figure 3.12, the signal $u(t)$ is applied to the left side of the line launching an incident wave of half that size, so that in all three cases

$$V'_{+}(s) = \frac{1}{2s} \tag{3.44}$$

This wave experiences an overall delay of T seconds in all cases before it arrives at the load. While in transit, when a signal arrives at an interface, some of the signal is reflected and some is transmitted. It is the transmission coefficient that determines how much of the signal is passed on to the right side of any interface. To get a better grasp of the progression of the signal to the load, for the three cases depicted in figure 3.12, we can refer to the propagation diagram of figure 3.13.

We are now ready to perform the requisite analysis on the circuit of figure 3.12a. This diagram is identical to figure 3.2. We will simply reproduce here, for our convenience, the parts of the solution needed for this analysis. From (3.4) we write the time constant at the load

$$t_s = CZ_0 \tag{3.45}$$

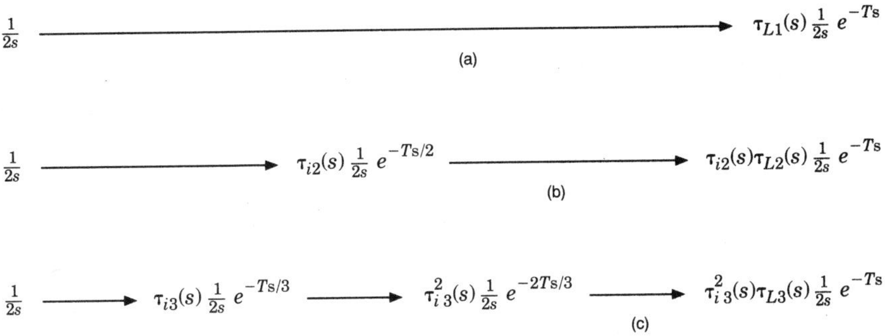

$\frac{1}{2s}$ ———————→ $\tau_{L1}(s)\,\frac{1}{2s}\,e^{-Ts}$

(a)

$\frac{1}{2s}$ ———→ $\tau_{i2}(s)\,\frac{1}{2s}\,e^{-Ts/2}$ ———→ $\tau_{i2}(s)\tau_{L2}(s)\,\frac{1}{2s}\,e^{-Ts}$

(b)

$\frac{1}{2s}$ ——→ $\tau_{i3}(s)\,\frac{1}{2s}\,e^{-Ts/3}$ ——→ $\tau_{i3}^{2}(s)\,\frac{1}{2s}\,e^{-2Ts/3}$ ——→ $\tau_{i3}^{2}(s)\tau_{L3}(s)\,\frac{1}{2s}\,e^{-Ts}$

(c)

Figure 3.13 A summary of the signal propagation to the load for the circuits of figure 3.12.

From (3.6) we write

$$\tau_{L1}(s) = \frac{2/t_s}{(s + 1/t_s)} \tag{3.46}$$

To find the expression for the voltage at the load of figure 3.12a, we multiply (3.46) by (3.44) and add the required e^{-Ts} to obtain

$$V_{s1}(s) = \frac{1/t_s}{s(s + 1/t_s)} e^{-Ts} \tag{3.47}$$

There is no time restriction on the solution represented by the above equation. The circuit shown in figure 3.12a is series matched at the source, hence any signal that is reflected from the load is absorbed at the source and there are no additional signals arriving at the load.

We now turn our attention to figure 3.12b. At the center of this circuit, the capacitance of value $C/2$ is shunted by a resistance Z_0 on each side; hence the time constant is $CZ_0/4$. Comparing this with (3.45) we conclude that the time constant has a value $t_s/4$. A voltage wave arriving from the left at this capacitor located at the center of the line sees the transmission line to the right of the capacitor as a resistance of value Z_0. The model that is representative of this situation is that shown in figure 3.8. We therefore use (3.32), with the time constant t_p replaced by $t_s/4$, to obtain

$$\tau_{i2}(s) = \frac{4/t_s}{(s + 4/t_s)} \tag{3.48}$$

At the load end of figure 3.12b we observe that the time constant is $CZ_0/2$ or $t_s/2$, hence, by direct reference to (3.46), we write the expression

$$\tau_{L2}(s) = \frac{4/t_s}{(s + 2/t_s)} \tag{3.49}$$

The expression for the voltage at the load of figure 3.12b is obtained by multiplying (3.49), (3.48), and (3.44), adding the required e^{-Ts}, to obtain

$$V_{s2}(s) = \frac{8}{t_s^2} \cdot \frac{1}{s(s + 4/t_s)(s + 2/t_s)} e^{-Ts}, \qquad \text{for } t < 2T \tag{3.50}$$

We are interested in dealing only with the response to the first wave which arrives at the load at time T. Some of this signal is reflected from the load and then travels in the reverse direction to the interface at the center of the line. It gets there at time $3T/2$, gets rereflected, and comes back to the load at time $2T$. To show that this newly arrived wave is excluded from the solution, we impose on the last equation the restriction $t < 2T$.

The procedure that was used to find the last expression is repeated for the case depicted in figure 3.12c. We state, without further explanation, that the transmission coefficient at the intermediate points on the transmission line is

$$\tau_{i3}(s) = \frac{6/t_s}{(s + 6/t_s)} \tag{3.51}$$

and the transmission coefficient at the load is

$$\tau_{L3}(s) = \frac{6/t_s}{(s + 3/t_s)} \tag{3.52}$$

The expression for the voltage at the load of figure 3.12c is obtained by multiplying (3.44) and (3.52) by the square of (3.51), as indicated in figure 3.13c, adding the required e^{-Ts}, to obtain

$$V_{s3}(s) = \frac{108}{t_s^3} \cdot \frac{1}{s(s + 3/t_s)(s + 6/t_s)^2} e^{-Ts}, \qquad \text{for } t < 5T/3 \tag{3.53}$$

Now that we have (3.47), (3.50), and (3.53), which are the expressions for the Laplace transforms of the output voltages for the circuits of figure 3.12, we are ready to proceed to find the corresponding time functions. Expanding the above three equations into partial fraction expansion form, we obtain

$$V_{s1}(s) = \left[\frac{1}{s} - \frac{1}{s + 1/t_s}\right]e^{-Ts} \tag{3.54}$$

$$V_{s2}(s) = \left[\frac{1}{s} - \frac{2}{s + 2/t_s} + \frac{1}{s + 4/t_s}\right]e^{-Ts}, \qquad \text{for } t < 2T \tag{3.55}$$

$$V_{s3}(s) = \left[\frac{1}{s} - \frac{4}{s + 3/t_s} + \frac{3}{s + 6/t_s} + \frac{6/t_s}{(s + 6/t_s)^2}\right]e^{-Ts}, \qquad \text{for } t < 5T/3 \tag{3.56}$$

The above three expressions can be readily inverted to obtain

$$v_{s1}(t) = \left[1 - e^{-(t-T)/t_s}\right]u(t - T) \tag{3.57}$$

$$v_{s2}(t) = \left[1 - 2e^{-2(t-T)/t_s} + e^{-4(t-T)/t_s}\right]u(t - T), \qquad \text{for } t < 2T \tag{3.58}$$

$$v_{s3}(t) = \left\{1 - 4e^{-3(t-T)/t_s} + 3\left[1 + 2\frac{t - T}{t_s}\right]e^{-6(t-T)/t_s}\right\}u(t - T), \qquad \text{for } t < 5T/3 \tag{3.59}$$

The above solutions represent the response of the three circuits shown in figure 3.12 to a step voltage input. It is again reiterated that the above solutions represent only the response to the first incident wave at the load. This presupposes that the second incident wave does not arrive at the load before the load voltage has reached its 50% point.

To get an idea of how the circuits shown in figure 3.12 behave, the results of the above equations are plotted in figure 3.14. The parameters that were chosen were $Z_0 = 50\,\Omega$, $C = 8\,\text{pF}$, and $T = 1\,\text{ns}$. We can see from the curves that as n increases, namely as the load capacitance is subdivided into smaller pieces and distributed more uniformly along the transmission line, that the steepness of the waveform at the load improves. This gives us reason to think that in the limit, as the load capacitor is distributed uniformly, there may

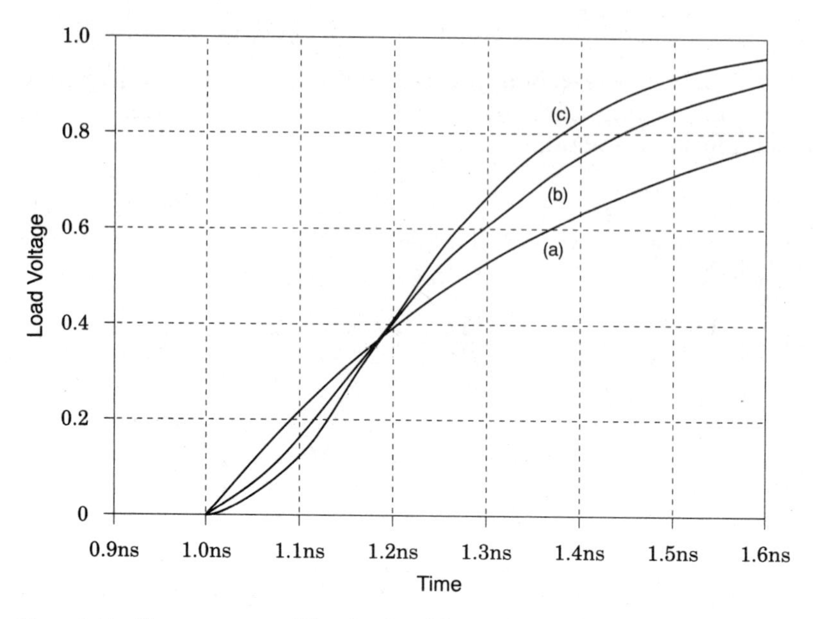

Figure 3.14 Step responses of the circuits of figure 3.12 with the parameters $Z_0 = 50\,\Omega$, $C = 8\,\text{pF}$, and $T = 1\,\text{ns}$.

be no waveform distortion at all. This question will be addressed in a later section.

3.5 Distributed Capacitive Load — Parallel Match

In the previous section we examined how distributing a capacitive load affects the behavior of a series-terminated transmission line. We wish to see if a parallel-terminated transmission line will exhibit similar behavior. We address the question with the circuits shown in figure 3.15. Figure 3.15a shows a parallel-terminated transmission line with the entire capacitor at the load. In figure 3.15b the capacitor has been split into two parts and distributed between equal lengths of the line. The process has been carried a step further in figure 3.15c where the capacitor has been divided into three parts and distributed uniformly.

As in the previous section, we will restrict our attention to the solution applicable to the arrival of the first incident wave. In figure 3.15a, the capacitor at the load is shunted on each side by a resistance Z_0, hence the time constant at the load is given by

$$t_{p1} = C\frac{Z_0}{2} \tag{3.60}$$

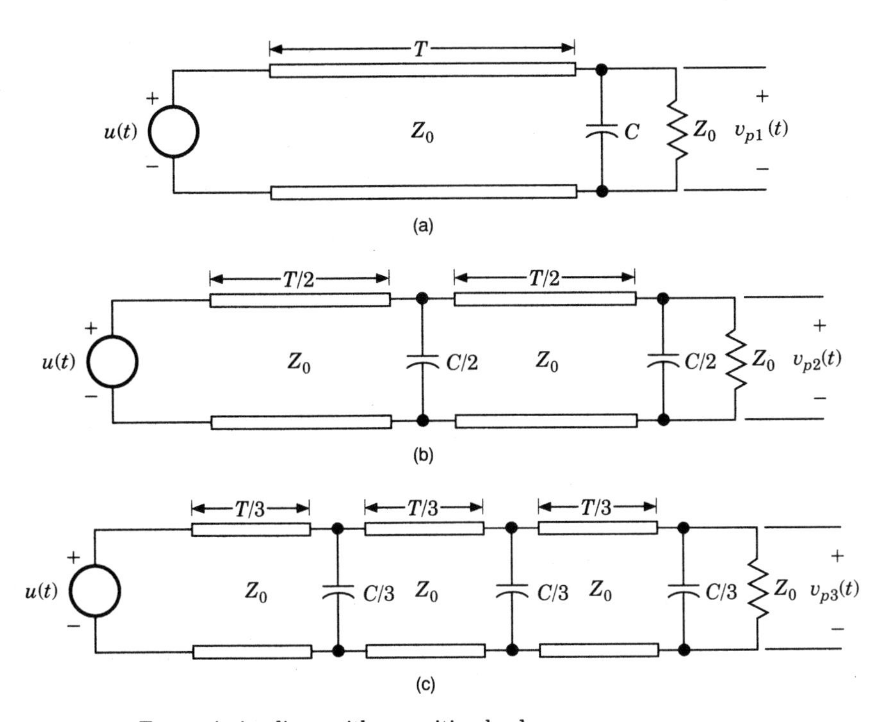

Figure 3.15 Transmission lines with capacitive loads.

In figure 3.15b each capacitor of value $C/2$ is shunted on each side by a resistance Z_0. The time constant t_{p2} for the circuit of figure 3.15b, at the interfaces and at the load, is given by

$$t_{p2} = \frac{C}{2} \cdot \frac{Z_0}{2} = t_{p1}/2 \tag{3.61}$$

As in figure 3.15b, in figure 3.15c each capacitor of value $C/3$ is shunted on each side by a resistance Z_0. The time constant t_{p3}, which is applicable at all the interfaces and at the load, is given by

$$t_{p3} = \frac{C}{3} \cdot \frac{Z_0}{2} = t_{p1}/3 \tag{3.62}$$

We see that if the process shown in figure 3.15 were continued, then a line divided into n sections would have the time constant $t_{pn} = t_{p1}/n$ at all the interfaces and at the load. Using the time constants above, the expressions for the transmission coefficients for each case shown in figure 3.15 follow directly from (3.32). We can write one general expression for all cases

$$\tau_{pn}(s) = \frac{n/t_{p1}}{s + n/t_{p1}} \tag{3.63}$$

In each case depicted in figure 3.15, the signal $V_S(s)$ is introduced on the left side of the line. It experiences an overall delay of T seconds in all cases before it arrives at the load. While in transit, when a signal arrives at an interface, some of it is reflected and some is transmitted. It is the transmission coefficient that determines how much of the signal is passed on to the right side of any interface. The progress of the signal in figure 3.15 can be very easily followed in the propagation diagram of figure 3.16.

To keep the analysis simple, we wish to restrict our attention to the first incident wave only. We therefore must restrict the time duration over which the solutions are applicable. In figure 3.15a, there is a first reflection at the load at time T. The first reflection returns to the source at time $2T$ to be reflected again, and becomes the second incident wave at the load at time $3T$. In figure 3.15b the first reflection, which also takes place at time T, requires $T/2$ seconds to get back to the first capacitive interface and comes back to the load at time $2T$. In figure 3.15c the first reflection requires $T/3$ seconds to get back to the first capacitive interface and comes back to the load at time $T + 2T/3$. In general, for a transmission line divided into n sections, the second reflected wave would arrive at time $(1 + 2/n)T$. The equations for the load voltages for the three circuits, along with proper time constraints, are

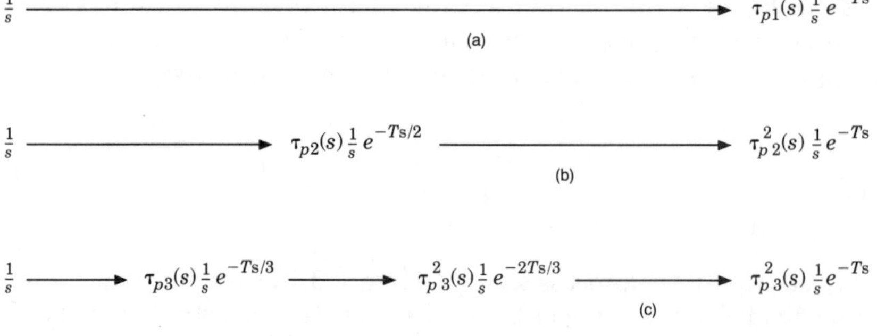

Figure 3.16 A summary of the signal propagation to the load for the circuits of figure 3.15.

$$V_{p1}(s) = \frac{1}{s}\tau_{p1}(s)e^{-Ts}, \qquad \text{for } t < 3T \tag{3.64}$$

$$V_{p2}(s) = \frac{1}{s}\,\tau_{p2}^{2}(s)e^{-Ts}, \qquad \text{for } t < 2T \tag{3.65}$$

$$V_{p3}(s) = \frac{1}{s}\,\tau_{p3}^{3}(s)e^{-Ts}, \qquad \text{for } t < 5T/3 \tag{3.66}$$

The above are for a step function input. We substitute (3.63) into the above equations to obtain

$$V_{p1}(s) = \frac{1/t_{p1}}{s(s + 1/t_{p1})}\,e^{-Ts}, \qquad \text{for } t < 3T \tag{3.67}$$

$$V_{p2}(s) = \frac{(2/t_p)^2}{s(s + 2/t_{p1})^2}\,e^{-Ts}, \qquad \text{for } t < 2T \tag{3.68}$$

$$V_{p3}(s) = \frac{(3/t_{p1})^3}{s(s + 3/t_{p1})^3}\,e^{-Ts}, \qquad \text{for } t < 5T/3 \tag{3.69}$$

Expanding the above into partial fraction form we have

$$V_{p1}(s) = \left[\frac{1}{s} - \frac{1}{s + 1/t_{p1}}\right]e^{-Ts}, \qquad \text{for } t < 3T \tag{3.70}$$

$$V_{p2}(s) = \left[\frac{1}{s} - \frac{1}{s + 2/t_{p1}} - \frac{2/t_{p1}}{(s + 2/t_{p1})^2}\right]e^{-Ts}, \qquad \text{for } t < 2T \tag{3.71}$$

$$V_{p3}(s) = \left[\frac{1}{s} - \frac{1}{s + 3/t_{p1}} - \frac{3/t_{p1}}{(s + 3/t_{p1})^2} - \frac{(3/t_{p1})^2}{(s + 3/t_{p1})^3}\right]e^{-Ts}, \qquad \text{for } t < 5T/3 \tag{3.72}$$

From the last equations we readily obtain the time domain expressions

$$v_{p1}(t) = \left[1 - e^{-(t-T)/t_{p1}}\right]u(t - T), \qquad \text{for } t < 3T \tag{3.73}$$

$$v_{p2}(t) = \left\{1 - \left[1 + \frac{t - T}{t_{p1}/2}\right]e^{-2(t-T)/t_{p1}}\right\}u(t - T), \qquad \text{for } t < 2T \tag{3.74}$$

$$v_{p3}(t) = \left\{1 - \left[1 + \frac{t - T}{t_{p1}/3} + \frac{1}{2}\left(\frac{t - T}{t_{p1}/3}\right)^2\right]e^{-3(t-T)/t_{p1}}\right\}u(t - T), \qquad \text{for } t < 5T/3 \tag{3.75}$$

The above solutions represent the response of the three circuits shown in figure 3.15 to a step voltage input. The results of the above equations are plotted in figure 3.17 for the same parameters as those used in the previous section, namely, $Z_0 = 50\,\Omega$, $C = 8\,\text{pF}$, and $T = 1\,\text{ns}$. We observe from the curves that the time at which $v_{pn}(t)$ reaches the 50% point moves closer to $t = T + t_{p1}$ as n gets larger. As was discussed above equation (3.64), the second incident voltage arrives at the load at time $T(1 + 2/n)$. For the above solutions to be applicable the second incident wave must not arrive before the load voltage has reached its 50% point. We therefore require that for the above solutions to be valid we must have

$$T + t_{p1} \le T(1 + 2/n) \tag{3.76}$$

or

$$t_{p1} \le 2T/n \tag{3.77}$$

In spite of the restrictions that exist in connection with (3.77), it is readily apparent that the step response improves as the capacitance is distributed in a uniform manner along the length of the line. We can foresee the possibility that we may be able to show that as C is

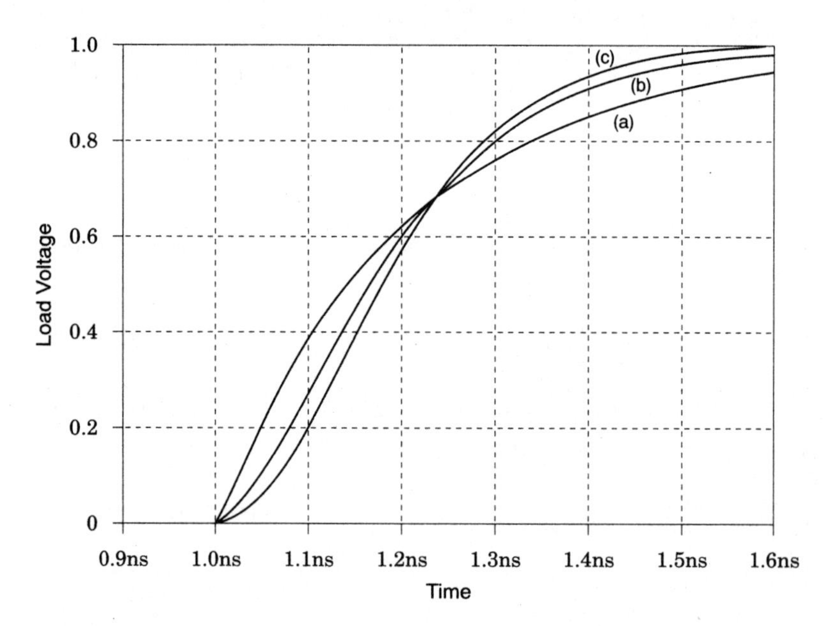

Figure 3.17 Step responses of the circuits of figure 3.15 with the parameters $Z_0 = 50\,\Omega$, $C = 8\,\text{pF}$, and $T = 1\,\text{ns}$.

divided into infinitesimal pieces and distributed along the length of the transmission line, that the input step might be transmitted in a distortionless manner.

3.6 Uniform Capacitive Loading — In the Limit

We are now ready to consider the question of dividing the load capacitance C into infinitesimal pieces for uniform placement along the transmission line. But before proceeding we first need to do some preliminary groundwork.

Digression — Students studying circuit theory are often posed the riddle of finding the input impedance R_{ab} for the infinite ladder structure shown in figure 3.18. All sections of the ladder structure are identical. A typical section is indicated between terminals a–b and terminals c–d consisting of the resistors R_1 and R_2.

The input impedance looking into terminals a–b is designated R_{ab}. If the first (leftmost) section of the ladder were removed, then the input impedance of the remaining ladder, R_{cd}, is equal to R_{ab}. This is because in an infinitely long structure one section more or less will not change the input impedance. For purposes of solving the problem, we can replace the circuit of figure 3.18 with the one shown in figure 3.19.

Working from the equivalent circuit we write directly

$$R_{ab} = R_1 + \frac{R_2 R_{ab}}{R_2 + R_{ab}} \tag{3.78}$$

The above leads to the quadratic equation in R_{ab}

$$R_{ab}^2 - R_1 R_{ab} - R_1 R_2 = 0 \tag{3.79}$$

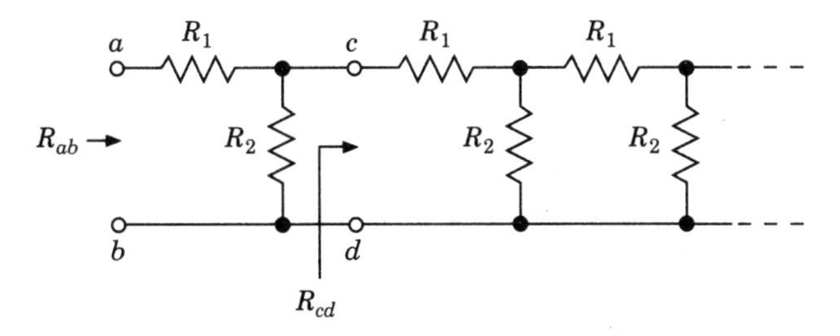

Figure 3.18 An infinite ladder structure.

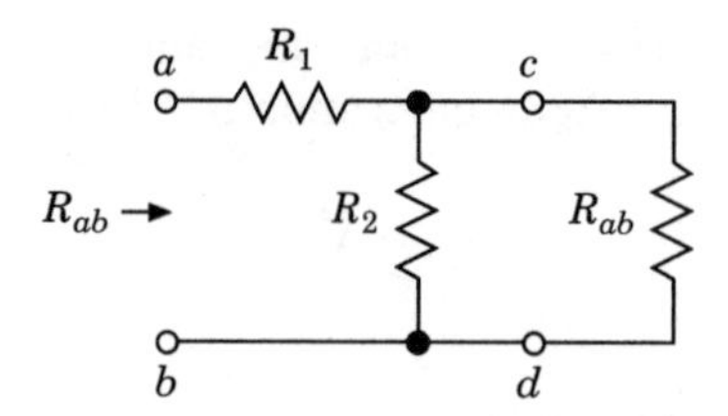

Figure 3.19 An equivalent circuit for the infinite ladder structure.

The only solution that produces positive results is

$$R_{ab} = \frac{R_1}{2}\left(1 + \sqrt{1 + 4\frac{R_2}{R_1}}\right) \tag{3.80}$$

This digression should help to understand the material that follows. ■

Consider figure 3.20, which shows two points on a distortionless line, one at position z and the other at position $z + l$. Assume that at position z we have an incident voltage $V_+(z, s)$. When this signal propagates to position $z + l$ it is delayed by T seconds, hence it takes the form

$$V_+(z + l, s) = V_+(z, s)e^{-Ts} \tag{3.81}$$

Now we will assume that at position z we have a reflected voltage $V_-(z, s)$. This voltage was at position $z + l$, T seconds earlier. Hence at $z + l$ it is described by

$$V_-(z + l, s) = V_-(z, s)e^{+Ts} \tag{3.82}$$

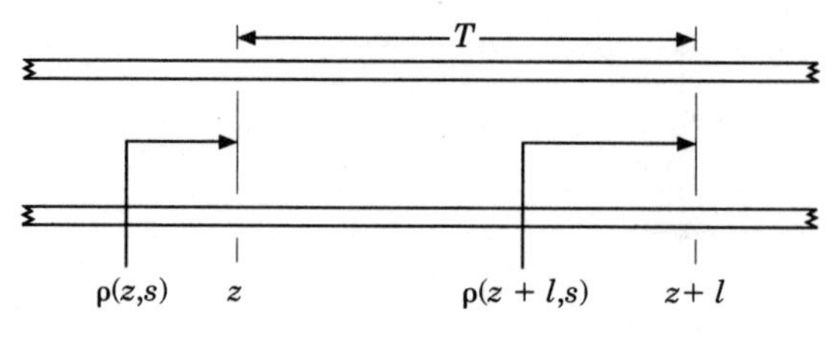

Figure 3.20 Transmission line model for finding the dependence of the reflection coefficient on z.

The reflection coefficient at position z is given by the ratio of reflected and incident waves at that position.

$$\rho(z, s) = \frac{V_-(z, s)}{V_+(z, s)} \tag{3.83}$$

Similarly the reflection coefficient at position $z + l$ is given by the ratio of the reflected wave of (3.82) to the incident wave of (3.81)

$$\rho(z + l, s) = \frac{V_-(z, s)}{V_+(z, s)} e^{2Ts} \tag{3.84}$$

Substituting (3.83) into the above we obtain the relationship

$$\rho(z + l, s) = \rho(z, s) e^{2Ts} \tag{3.85}$$

We will now use this expression to help us obtain the ratio of total voltage to total current at any point on the line. The result will naturally be the impedance seen looking into a transmission line. For convenience we rewrite (3.2) in terms of the notation used in this section

$$\rho(z, s) = \frac{Z_L(z, s) - Z_0}{Z_L(z, s) + Z_0} \tag{3.86}$$

The term $Z_L(z, s)$ is the impedance seen by an observer standing at position z on the line and looking to the right. The above can be recast in terms of admittances, which are the reciprocals of impedances, as

$$\rho(z, s) = \frac{Y_0 - Y_L(z, s)}{Y_0 + Y_L(z, s)} \tag{3.87}$$

An observer located at some position z, looking to the right, sees a reflection coefficient $\rho(z, s)$. This information can be used in (3.87) to find the admittance seen to the right of the observer. Solving for the explicit relation, we get

$$Y_L(z, s) = \frac{1 - \rho(z, s)}{1 + \rho(z, s)} Y_0 \tag{3.88}$$

With the preliminaries out of the way we can now turn our attention to figure 3.21 which shows cascaded sections of a transmission line.

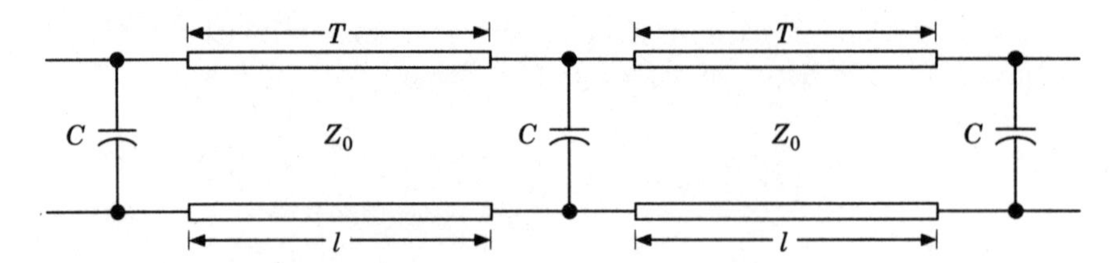

Figure 3.21 Infinitely long cascaded lines.

Each section of the line has a one-way signal propagation delay of T seconds. At each transmission-line interface we have a capacitance C shunting the line.

In figure 3.22 we have taken each section of the cascaded lines of figure 3.21 and divided it into n subsections. As a consequence each section of the new transmission line has a one-way signal propagation delay of T/n seconds, and at each transmission-line interface we have a capacitance C/n shunting the line. We apply the same argument that was used in the digression at the beginning of this section. Since the cascaded line is infinitely long, then it is reasonable to assume that the impedance seen looking to the right into any section of the transmission line is the same. We will refer to this as the *modified characteristic impedance* Z_{0M}.

The admittance of the capacitor whose value is C/n is

$$Y_{Cn} = sC/n \tag{3.89}$$

The admittance seen by an observer standing to the left of b—b′ and looking to the right is

$$Y_L(s) = Y_{0M} + sC/n \tag{3.90}$$

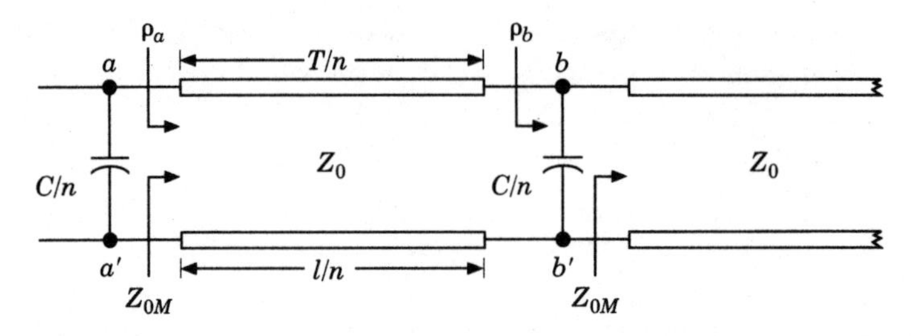

Figure 3.22 Cascaded lines going to the limit.

The reflection coefficient ρ_b, indicated in figure 3.22, is found by substituting (3.90) into (3.87), with the result

$$\rho_b(s) = \frac{Y_0 - Y_{0M} - sC/n}{Y_0 + Y_{0M} + sC/n} \tag{3.91}$$

We now use (3.85) to find the reflection coefficient ρ_a indicated in figure 3.22.

$$\rho_a(s) = \frac{Y_0 - Y_{0M} - sC/n}{Y_0 + Y_{0M} + sC/n} e^{-2Ts/n} \tag{3.92}$$

If the transmission line is infinitely long, then it is fair to assume that the impedance seen looking into any section of the transmission line is the same as for any other section of the line. We therefore substitute (3.92) into (3.88) and set the resultant expression for admittance to Y_{0M} to get

$$Y_{0M} = \frac{1 - \dfrac{Y_0 - Y_{0M} - sC/n}{Y_0 + Y_{0M} + sC/n} e^{-2Ts/n}}{1 + \dfrac{Y_0 - Y_{0M} - sC/n}{Y_0 + Y_{0M} + sC/n} e^{-2Ts/n}} Y_0 \tag{3.93}$$

The modified characteristic admittance Y_{0M} appears on both sides of (3.93). We solve for Y_{0M} to obtain the quadratic equation

$$Y_{0M}^2 + (sC/n)Y_{0M} - Y_0^2 - Y_0(sC/n)\frac{1 + e^{-2Ts/n}}{1 - e^{-2Ts/n}} = 0 \tag{3.94}$$

As n approaches infinity, it is clear that the second term on the left side of the above equation vanishes. We can also show that

$$\lim_{n \to \infty}(sC/n)\frac{1 + e^{-2Ts/n}}{1 - e^{-2Ts/n}} = \frac{C}{T} \tag{3.95}$$

so that (3.94) can be reduced to

$$Y_{0M}^2 - Y_0^2\left(1 + \frac{C}{TY_0}\right) = 0 \tag{3.96}$$

which is solved by

$$Y_{0M} = Y_0\sqrt{1 + \frac{C}{TY_0}} \tag{3.97}$$

From (1.51) and (1.52) we readily obtain

$$Y_0 = \sqrt{\frac{C_0}{L_0}} \tag{3.98}$$

and

$$T = \frac{l}{\nu} = l\sqrt{C_0 L_0} \tag{3.99}$$

We use the above two equations in (3.97) to obtain

$$Y_{0M} = \sqrt{\frac{C_0 + C/l}{L_0}} \tag{3.100}$$

The reciprocal of the above is

$$Z_{0M} = \sqrt{\frac{L_0}{C_0 + C/l}} \tag{3.101}$$

From (1.51) and (1.52) we can readily obtain the relationship

$$\nu = \frac{Z_0}{L_0} \tag{3.102}$$

In our case we expect that the speed of propagation will be modified by the capacitive loading of the line, so in place of (3.102) we will use

$$\nu_M = \frac{Z_{0M}}{L_0} \tag{3.103}$$

Applying (3.103) to (3.101) we conclude that ν_M, the modified speed of propagation on the line, is

$$\nu_M = \frac{1}{\sqrt{L_0(C_0 + C/l)}} \tag{3.104}$$

We see in (3.101) and (3.104) that the characteristic impedance of the line and the speed of propagation are determined by the distributed inductance L_0 and by the distributed capacitance C_0 which has been augmented by the capacitance C/l.

The above results can be summarized in the following (more convenient) expressions relating modified transmission-line parameters to the original ones. The modified characteristic impedance is reduced by distributed capacitive loading

$$Z_{0M} = \frac{Z_0}{\sqrt{1 + C/lC_0}} \tag{3.105}$$

as is the modified speed of propagation

$$\nu_M = \frac{\nu}{\sqrt{1 + C/lC_0}} \tag{3.106}$$

Since the length of the line and the speed of propagation determine the one-way delay, it follows that the one-way delay is increased by the distributed capacitive loading.

$$T_M = T\sqrt{1 + C/lC_0} \tag{3.107}$$

Example 3.1 A 72 Ω transmission line with a one-way delay of 5 ns is connected, as shown in figure 3.23a. The line is 1.1 m in length. Five 12 pF capacitors are connected at the load end of the line so that the capacitance C has a value of 60 pF. The step response of this circuit is shown in figure 3.24a. It has the characteristic exponential rise to a steady-state value of unity.

To see if there is indeed an improvement in the step response at the load if the load capacitance is divided into five parts and distributed uniformly along the line, we investigate the circuit of figure 3.23b. The step response, obtained using PSpice*, is shown in figure 3.24b. Although this circuit has a much steeper rise time, it has a very substantial overshoot. From the given data we use (3.105) to find that the modified characteristic impedance is now

$$Z_{0M} = 53\,\Omega$$

The above-mentioned overshoot may be due to the fact that the 72 Ω load resistor no longer matches the line which has a modified characteristic impedance of 53 Ω. A 72 Ω load on a 53 Ω line produces a positive reflection coefficient. To correct the problem, the 72 Ω terminating resistor in figure 3.23b is replaced by the 53 Ω load resistor shown in 3.23c. The resultant step response in figure 3.24c shows that the overshoot has been substantially reduced and that the rise time has hardly changed.

*PSpice is a registered trademark of MicroSim Corporation.

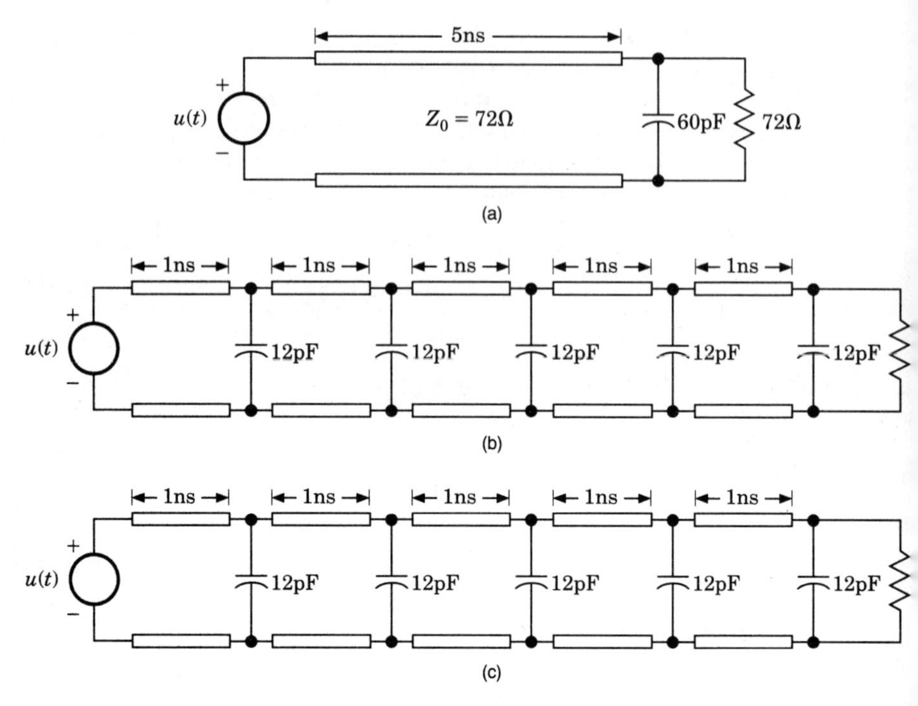

Figure 3.23 Three circuit configurations for testing the theory of this section.

The above conclusion is reinforced when we use (3.107) to find that

$$T_M = 6.8\,\text{ns}$$

Inspection of figure 3.24 shows that the 50% point delay is very close to the calculated value of 6.8 ns. We can conclude that even when the load capacitor is divided into only five parts, as was done for the circuit of figure 3.23, the theory of uniform capacitive loading produces reasonably accurate results. ■

3.7 Conclusion

In this chapter it was desired to demonstrate that distributing capacitive loads is superior to connecting them all at one location. To do this we showed in sections 3.2 and 3.3 that placing a capacitive load at the end of a transmission line causes rise-time degradation in the step response as well as in the response to a unit step with ramp transition. This was carried out for both series and parallel matched lines in order to show that the method of termination has no effect on the issue at hand. In sections 3.4 and 3.5 we showed that subdividing a capacitive load and distributing it uniformly along the transmission line shows potential for improvement of the step response. Again this was done

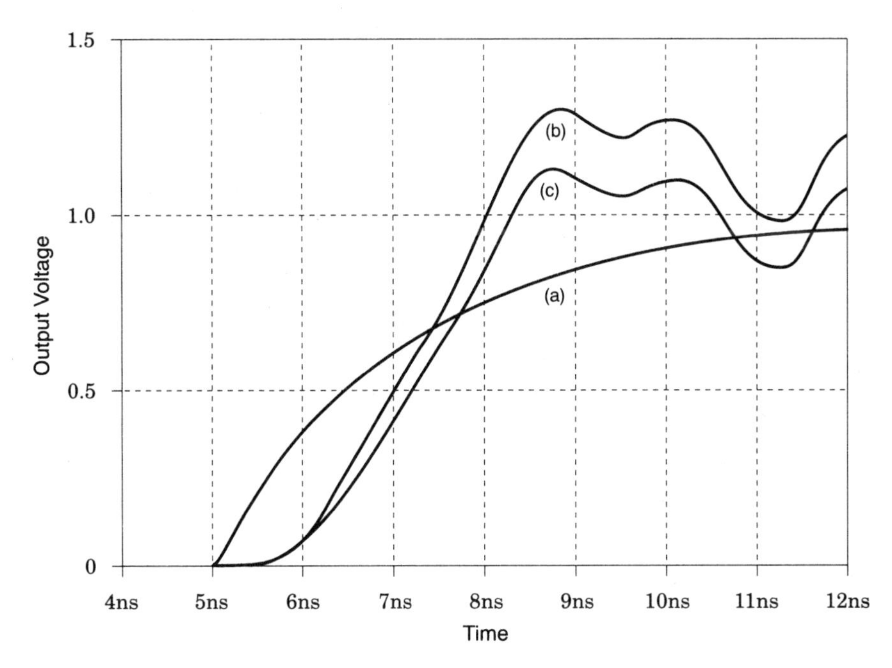

Figure 3.24 The load voltages of the circuits of figure 3.23 as evaluated by the use of PSpice.

for both series or parallel terminated lines to show that the above is true irrespective of the type of termination used. In section 3.6 it was shown that taking the capacitance and distributing it uniformly along the transmission line causes a change in the characteristic impedance of the line and also affects the speed of propagation of the signals along the line. Aside from the above two changes, however, the line remains distortionless. This analysis was performed for infinite lines, hence the result is independent of the type of termination used.

In summary, the above analysis allows us to draw the conclusion that, where possible, it is better to distribute logic-device gate-inputs in a uniform manner along the length of the line rather than lump them at the load end of a transmission line.

Problems

P3.1 Using a computer circuit analysis program such as PSpice, verify the curves shown in figure 3.4.

P3.2 In figure 3.2 replace the load with the devices indicated and carry out the Laplace transform analysis to produce a result similar to figure 3.4 for the following cases:

(a) The load is Z_0 connected in series with a capacitor C.
(b) The load is Z_0 connected in series with an inductor L.
(c) The load is Z_0 connected in parallel with an inductor L.

P3.3 In reference to the voltage $v_I(t)$ presented in (3.28) and figure 3.6, analytically find the instant of time at which the minimum beyond $t = 2T$ occurs and find its value.

P3.4 Using a computer circuit analysis program such as PSpice, verify the curves shown in figure 3.6. Note that the computer simulation produces results that are not as accurate as those obtained analytically.

P3.5 Using a circuit analysis program such as PSpice, verify the curves shown in figure 3.10.

P3.6 Using a circuit analysis program such as PSpice, verify the curves shown in figure 3.11.

P3.7 Verify the partial fraction expansions that take (3.67)–(3.69) into (3.70)–(3.72).

P3.8 Verify the connection between the time functions given in (3.73)–(3.75) with the Laplace transform expressions of (3.70)–(3.72).

P3.9 Using a circuit analysis program such as PSpice, verify the curves shown in figure 3.14.

P3.10 Using a circuit analysis program such as PSpice, verify the curves shown in figure 3.17.

P3.11 Equations (3.67)–(3.69) were obtained for a step function input. Note that the time restriction applicable to the above-mentioned equations is due to the fact that we restrict our attention to the solution applicable to the arrival of the first incident wave.

(a) Rewrite (3.67)–(3.69) for a delta function input and generalize as a function of the subscript n.
(b) Write the general time function for the above Laplace transform.
(c) Differentiate the above time function and determine the point at which the maximum occurs.
Solution: $t_{max} = T + [(n - 1)/n]t_{p1}$.
(d) Find the value at the peak and show that as n approaches infinity the peak goes to infinity. **Hint:** Stirling's formula for determining $n!$ for large n may be helpful.

$$n! \approx n^n e^{-n}\sqrt{2\pi n}, \qquad \text{for large } n$$

(e) By starting with the definition of the Laplace transform, show that the integral over all time of a time function can be determined by evaluating its Laplace transform at $s = 0$. Using this and the result of part (d), is it reasonable to say that the general function found in part (b) approaches a delta function as $n \to \infty$?

(f) In light of the above analysis, is it reasonable to say that as n approaches infinity, the line becomes distortionless with a fixed delay equal to $T + t_p$? What restriction stated at the beginning of this problem causes this analysis to be defective?

P3.12 Verify the claim made in (3.95) and as a consequence show that (3.94) leads to (3.96) as $n \to \infty$.

P3.13 We wish to verify the results given in example 3.1.

(a) Determine the time constant of the circuit in figure 3.23a and analytically verify the curve of figure 3.24a.

(b) Verify the results shown in figure 3.24 by the use of a computer circuit simulation program such as PSpice.

(c) Apply (3.105) and (3.107) to the given data to verify that the 53 Ω is appropriate for the circuit of figure 3.23c and that $T_M = 6.8$ ns as stated.

(d) Rerun part (b) for a longer time duration, for example 40 ns, to show that the three curves shown in 3.24 tend to a steady-state value of unity.

(e) Does the circuit of figure 3.23c demonstrate improved performance? Explain the initial overshoot in the step response of the circuit of figure 3.23b.

P3.14 A 120 Ω transmission line with a one-way delay of 2 ns is connected, as shown in figure 3.23a. The line is 0.4 m in length. Eight 3 pF capacitors are connected at the load end of the line.

(a) Analytically find and plot the step response of the line. What is the 50% point delay for this signal?

(b) Assume the eight capacitors have been uniformly distributed along the transmission line, as in figure 3.23c. Find the total signal delay and from that calculate the 50% point delay of the signal.

(c) It is advisable to change the load resistance of this line. Find the new value of load resistance.

P3.15 The problem treated in example 3.1 was for a shunt terminated line. It is desired to redo this example, using a circuit analysis program such as PSpice, for the series terminated case shown in figure 3.25.

(a) Obtain curves similar to those shown in figure 3.24 for the series terminated case.

(b) Redo the curves again for the range $4\,\text{ns} \le t \le 40\,\text{ns}$ to show that the steady-state voltage tends toward unity.

(c) Does the circuit of figure 3.25c demonstrate improved performance? Explain the initial undershoot in the step response of the circuit of figure 3.25b.

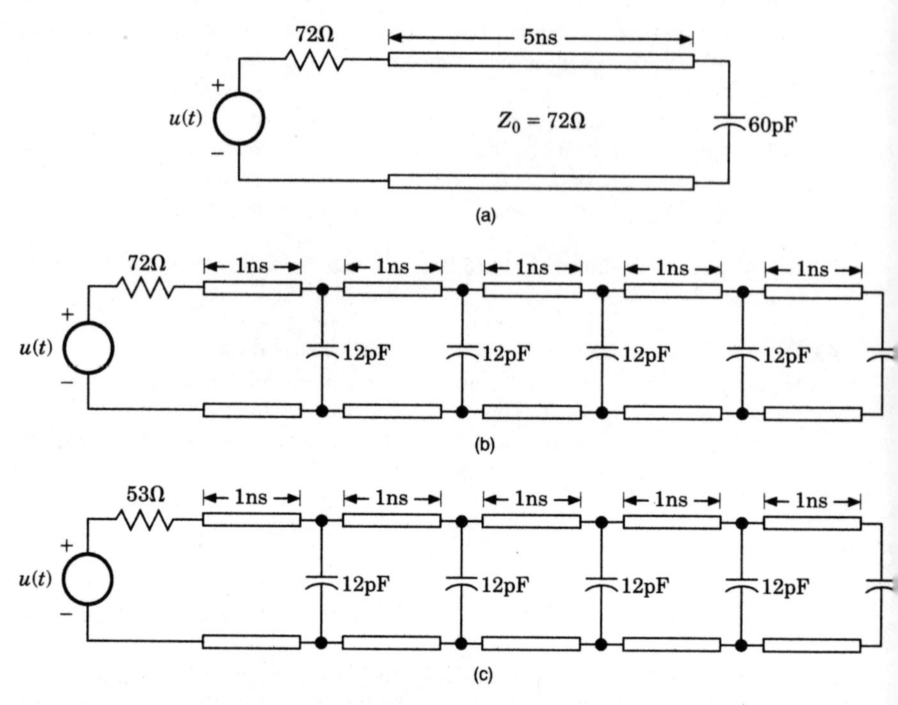

Figure 3.25 Three circuit configurations for the series terminated case.

P3.16 A 90 Ω transmission line of length 2 m has a speed of propagation of 0.8c. A capacitance of 60 pF is uniformly distributed along the length of the line.

(a) Find L_0 and C_0 for the existing line.

(b) Determine the modified Z_{0M} and ν_M for the line with the added capacitance.

Chapter

4

Non-linear Sources and Loads

4.1 Introduction

In dealing with reflections on transmission lines in chapter 2, we considered various transmission-line problems which were solved for linear sources and for linear loads. The driving gate was modeled as a voltage source in series with an output resistance and the receiving gate was represented as a resistor. For certain types of digital devices, most notably Transistor-Transistor Logic (TTL) as well as Complementary Metal Oxide Semiconductor (CMOS) logic, neither the driving-gate output nor the receiving-gate input can be represented in terms of linear models. The techniques developed in chapter 2 cannot be readily applied in those cases.

The load-line method of analysis* which is presented in the next section is useful for problems for which the linear method fails. The approach to presenting this method of solving the problem will be first to make the technique tangible by demonstrating it on linear models, then showing how it can be applied to non-linear models.

4.2 The Method of Load-Line Analysis

The method of load-line analysis is quite old. Anyone who has studied biasing of transistors (or electron tubes) has used this technique. Its application to transmission-line problems has been previously

*This is also referred to as the Bergeron-plot method [1,2].

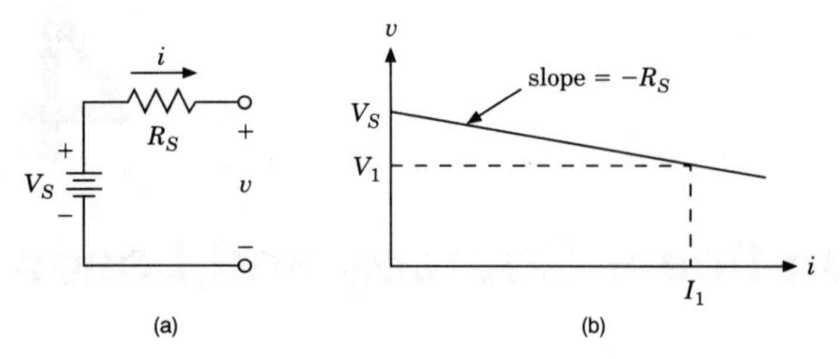

Figure 4.1 (a) A voltage source with output resistance R_S. (b) The associated voltage-current characteristic.

demonstrated [1–3]. This section will be dedicated to explaining the method. The sections that follow will show how this method can be easily applied to transmission-line problems.

The voltage source shown in figure 4.1a has the voltage-current characteristic shown in figure 4.1b. To obtain this result we start by writing the voltage-current relationship for the voltage source, which is

$$v = V_S - R_S\, i \tag{4.1}$$

The above describes a straight line with slope $-R_S$ shown in figure 4.1b. The higher the value of R_S, the steeper the slope of the voltage-current characteristic.

The resistive load shown in figure 4.2a has the voltage-current characteristic shown in figure 4.2b. This is merely a consequence of Ohm's law, which very obviously states that

$$v = R_L\, i \tag{4.2}$$

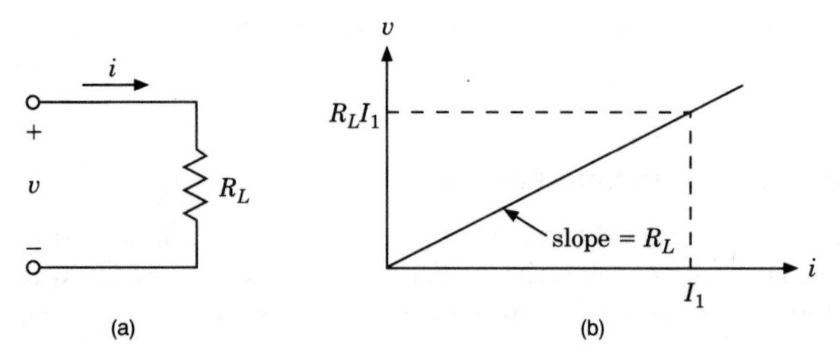

Figure 4.2 (a) A resistive load of value R_L. (b) The associated voltage-current characteristic.

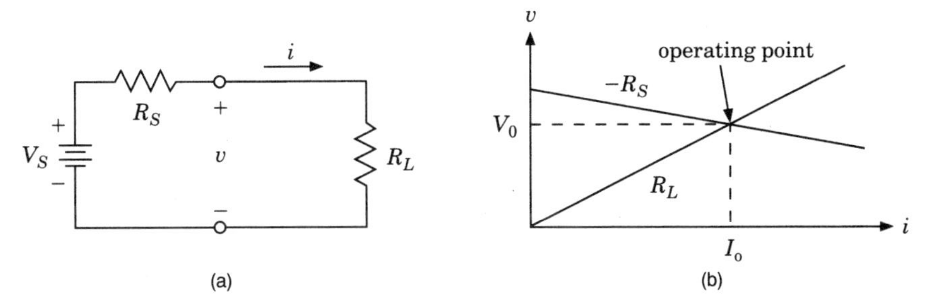

Figure 4.3 A solution for the circuit shown on the left by the graphical construction appearing on the right.

The circuit in figure 4.3a is composed of the elements shown in 4.1a and 4.2a. We could easily solve for the current i in figure 4.3a analytically. Since (4.1) and (4.2) constitute a set of simultaneous equations, we could simply substitute one into the other to find the solution that satisfies both. But we wish to lay the foundation for dealing with non-linear sources and loads, so we will proceed with the graphical method. It is based on the fact that in order to solve a set of simultaneous equations, we merely find the point of intersection of the two curves that the two equations describe. The resultant graphical load-line technique* is shown in figure 4.3b.

As has been mentioned previously, the above method is not particularly needed when dealing with linear problems. It can be put to good use for cases in which the source and load are non-linear, as may be the case for the two gates shown in figure 4.4a. Hypothetical source and load characteristics are drawn on the graph in 4.4b. (It is

*Henceforth, a value of resistance appearing next to a straight line will denote the slope of that line.

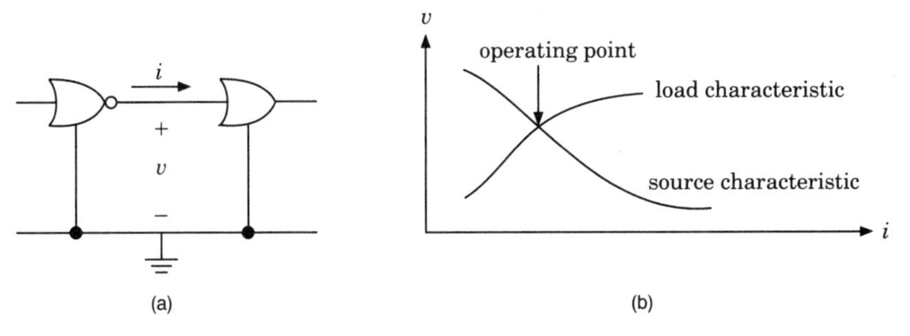

Figure 4.4 A solution for the non-linear circuit shown on the left by use of the graphical construction appearing on the right.

worth noting that the current i shown in that figure, which is used in the remainder of this book, has the opposite polarity of that used in gate specification-sheets published by manufacturers.) The point of intersection of the two curves is the operating point of the circuit.

The digital gate driving the left side of the circuit in figure 4.4a will invariably have more than one state. This is taken care of by the graph with the two source characteristics shown in 4.5. It is very important to keep in mind that the solution must always lie on the two curves that determine the circuit operation. When the circuit of figure 4.4a is in the low state, the solution must lie simultaneously on the low-state source-characteristic and on the load characteristic, as indicated in figure 4.5—similarly when the circuit is in the high state.

Example 4.1 The left gate shown in figure 4.4a has a non-linear output characteristic, which is purely hypothetical, given by

$$v = 2e^{-i} \quad \text{in the low state} \tag{4.3}$$

$$v = 5e^{-i} \quad \text{in the high state} \tag{4.4}$$

The load consists of the gate on the right which has a non-linear input characteristic, hypothetical as well, given by

$$v = 10\left[1 - e^{-0.6i}\right] \tag{4.5}$$

It is desired to solve for the voltage and current at the load by using analytical and graphical analyses.

By numerical trial and error it is found for the low state that (4.3) and (4.5) are solved by $v = 1.52\,\text{V}$ and $i = 0.275\,\text{A}$. Similarly it is found for the high state that (4.4) and (4.5) are solved by $v = 2.86\,\text{V}$ and $i = 0.560\,\text{A}$. For the graphical solution, the three curves of (4.3–4.5) are plotted in figure 4.6. It is clear that the solutions obtained graphically are identical to those obtained analytically, even if the graphical solution is not as accurate. ■

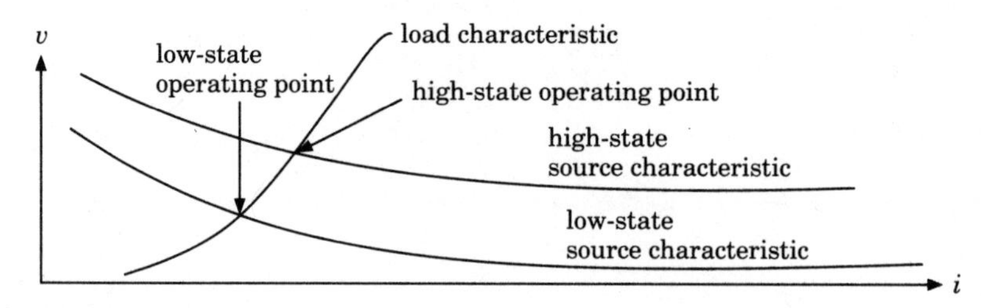

Figure 4.5 Two operating points found for a two-state non-linear gate driving a non-linear load.

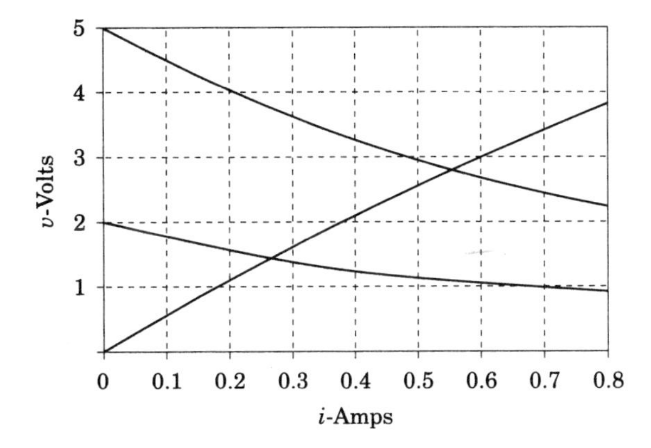

Figure 4.6 Graphical load-line analysis for example 4.1.

4.3 Transmission-Line Load-Line Analysis

We wish to extend the analysis of the previous section to the case of a source and a load that are separated by a transmission line, as shown in figure 4.7. The presence of the transmission line complicates the analysis somewhat, so some additional explanation will be required.

Assume that the entire circuit has reached steady state in the low state. For this condition the operating point for the entire line is found at the point (V_1, I_1) in figure 4.8. Suddenly the output of the left gate changes to the high-state source-characteristic and the operating point at the output gate must immediately change to this curve. Accordingly, an incident voltage wave V_{i1} and an incident current wave I_{i1} are launched from the left end of the transmission line and operation instantly shifts to the point (V_2, I_2), which lies on the high-state source-characteristic. From chapter 1 we know that the incident voltage and current waves are in the ratio of Z_0, hence the line connecting the operating point (V_1, I_1) to the operating point (V_2, I_2) has a slope $+Z_0$.

The incident voltage wave V_{i1}, and the accompanying incident current wave I_{i1}, propagate toward the right side load. Within the time duration T, which is the one-way time-delay for the transmission line, the voltage on the entire line becomes V_2 and the current becomes I_2. When the incident wave arrives at the receiving gate, which is

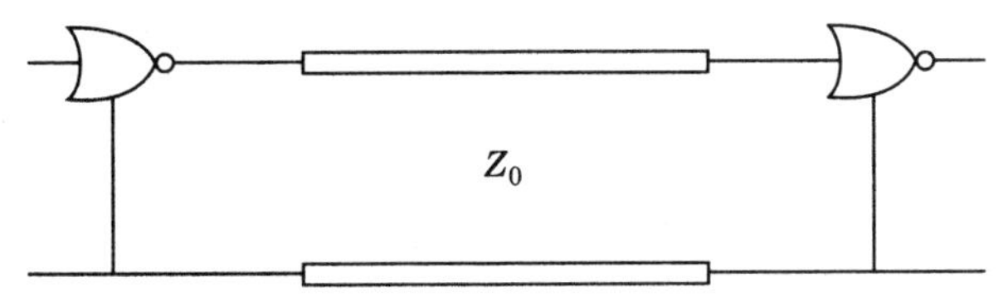

Figure 4.7 Two logic gates separated by a transmission line.

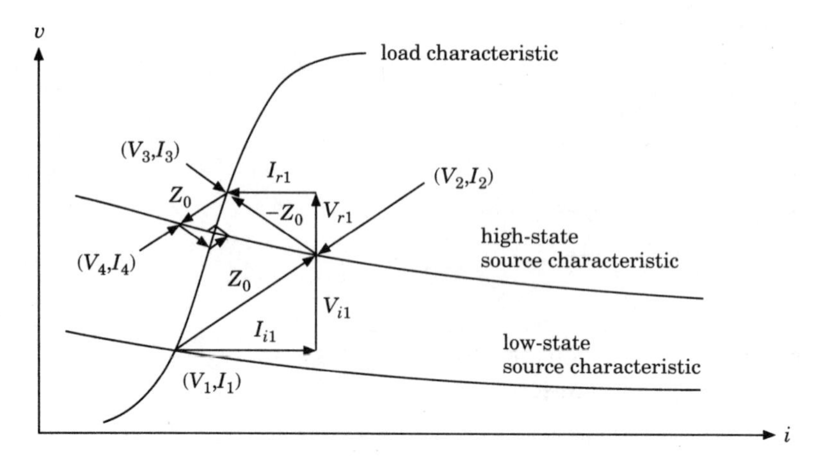

Figure 4.8 Load-line diagram for the two logic gates shown in figure 4.7.

the load, it is found that the point (V_2, I_2) does not lie on the load characteristic. Therefore a reflected voltage wave V_{r1}, accompanied by the reflected current wave I_{r1}, is launched at the load. But from chapter 1 we know that the reflected waves are in the ratio of $-Z_0$, hence the line connecting the operating point (V_2, I_2) to the operating point (V_3, I_3) has a slope $-Z_0$. This action is shown in figure 4.8.

As the reflected waves propagate from right to left, they establish the voltage V_3 and the current I_3 on the line. When the reflected waves arrive at the left (sending) end at time $2T$, it is found that the point (V_3, I_3) does not lie on the high-state source-characteristic, and new incident waves are launched to cause the operating point to change to (V_4, I_4). The entire sequence of events repeats until the diagram spirals in toward the steady state.

The work appearing on the load-line plot can be transferred to a lattice diagram for clarity, as shown in figure 4.9. To obtain this plot, the voltages V_1, V_2, V_3, and V_4, from the load-line diagram, are entered into the triangles in the voltage lattice diagram. From these data the values of the incident and reflected waves are readily found using

$$V_{i1} = V_2 - V_1 \tag{4.6}$$

$$V_{r1} = V_3 - V_2 \tag{4.7}$$

$$V_{i2} = V_4 - V_3 \tag{4.8}$$

In a similar manner the currents I_1, I_2, I_3, and I_4, from the load-line diagram, are entered into the triangles in the current lattice diagram.

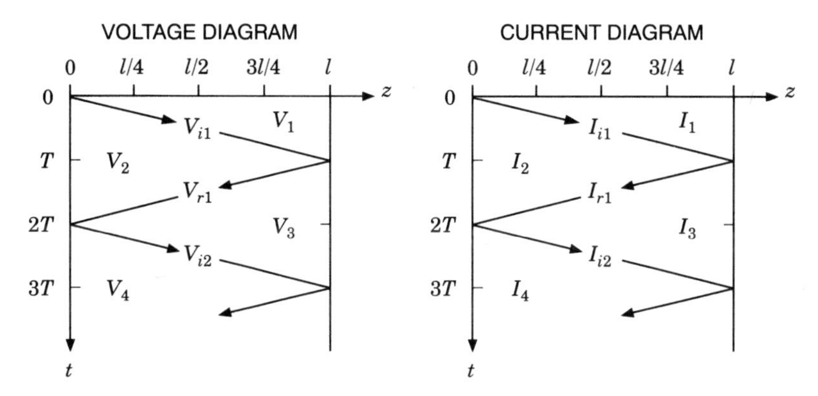

Figure 4.9 The load-line diagram of figure 4.8 reinterpreted into a lattice diagram.

From these data the values of the incident and reflected waves are readily found using

$$I_{i1} = I_2 - I_1 \tag{4.9}$$

$$I_{r1} = I_3 - I_2 \tag{4.10}$$

$$I_{i2} = I_4 - I_3 \tag{4.11}$$

An example will be used to demonstrate the method.

Example 4.2 The left gate shown in figure 4.7 has an output characteristic, purely hypothetical, given by

$$v = 2e^{-5i} \qquad \text{in the low state} \tag{4.12}$$

$$v = 5e^{-5i} \qquad \text{in the high state} \tag{4.13}$$

The right gate has a load characteristic, also purely hypothetical, given by

$$v = 6\left[1 - e^{-40i}\right] \tag{4.14}$$

The transmission line connecting the two gates has a characteristic impedance $Z_0 = 50\,\Omega$. The circuit is in the low-state steady-state. The left gate suddenly changes to the high state. It is desired to plot the voltage waveforms at the sending and receiving ends of the transmission line.

This problem is solved by using the load-line technique described in this section. After plotting the curves described in (4.12–4.14) it is found that the low-state steady-state is found at the point

$$(V_1, I_1) = (1.91\,\text{V},\ 9.6\,\text{ma})$$

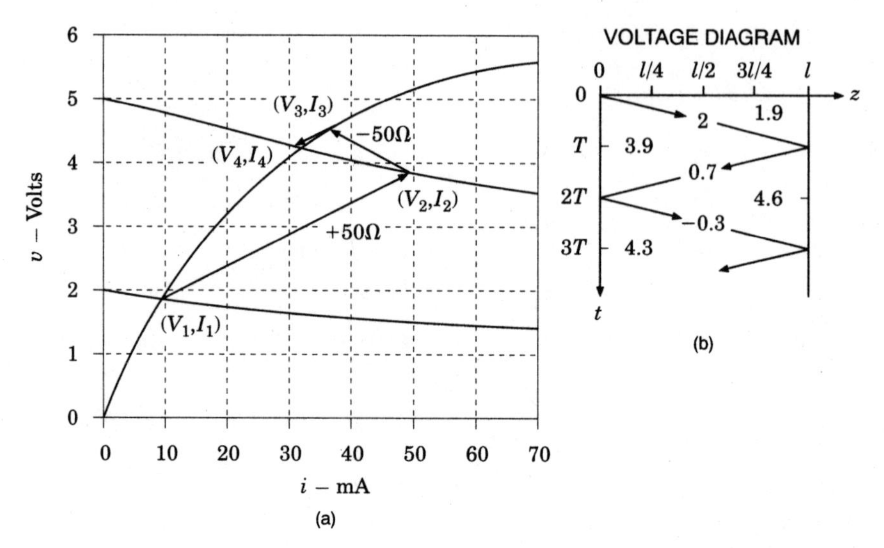

Figure 4.10 Load-line diagram (a) and lattice diagram (b) for the problem in example 4.2.

as shown in figure 4.10a. From this point we draw a line with a slope of $+50\,\Omega$ until this intersects the high-state source-characteristic. The point of intersection is at

$$(V_2, I_2) = (3.9\,\text{V}, 49.5\,\text{ma})$$

We continue from this point with a line of slope of $-50\,\Omega$ until this hits the load characteristic. This occurs at the point

$$(V_3, I_3) = (4.6\,\text{V}, 36\,\text{ma})$$

We now use a line with a slope of $+50\,\Omega$ to complete the plot to the point

$$(V_4, I_4) = (4.3\,\text{V}, 30.3\,\text{ma})$$

This point is so close to the high-state steady-state point

$$(V_{SS}, I_{SS}) = (4.28\,\text{V}, 31.2\,\text{ma})$$

that there is no additional interest in continuing the process.

The work appearing on the load-line plot of figure 4.10a was transferred to the lattice diagram shown in figure 4.10b. This in turn was used to plot the voltage waveforms at the source and load ends of the transmission line, as shown in figure 4.11. In fact, once the lattice diagram is drawn, the voltage waveforms can be plotted for any location on the line. ■

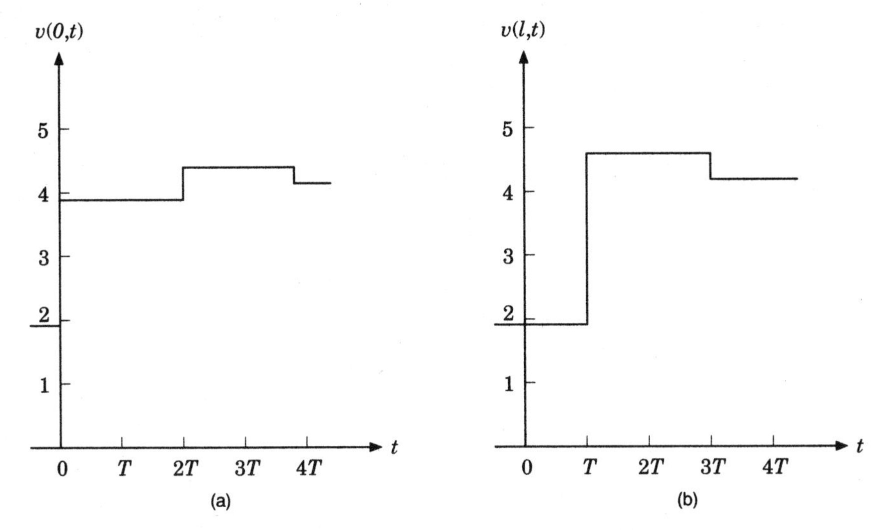

Figure 4.11 Voltage waveforms at the sending end (a) and receiving end (b) of the line of example 4.2.

Example 4.3 We have the same circuit as specified in example 4.2, but at the outset the circuit is in the high-state steady-state. The left gate suddenly changes to the low state. It is desired to plot the voltage waveforms at the center and receiving ends of the transmission line.

We follow the same procedure as we did in the previous example with the result shown in figure 4.12. It is noteworthy that the current becomes negative for a time duration. We will assume that the gates used in figure 4.7 allow

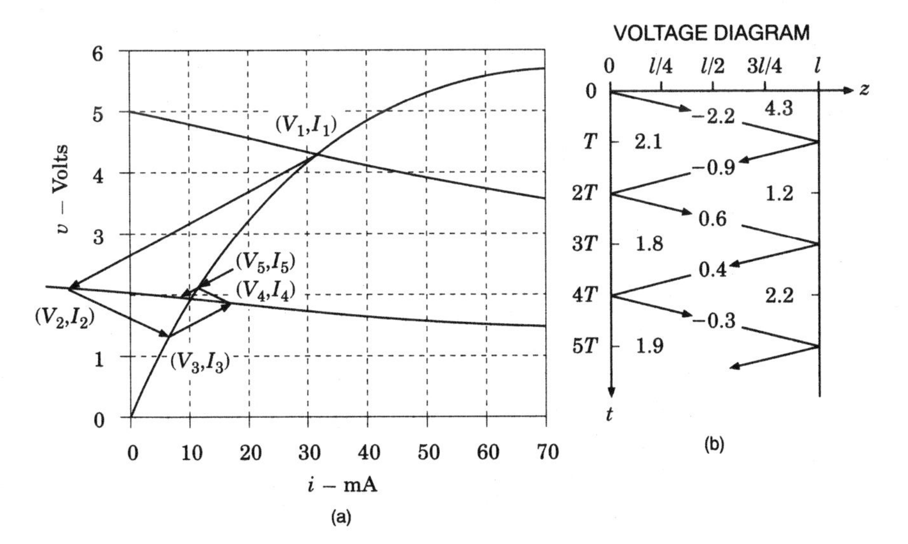

Figure 4.12 Load-line diagram (a) and lattice diagram (b) for the problem in example 4.3.

current to enter the gate output when the gate is in the low state, as would definitely be the case for TTL and CMOS. The voltage waveforms at the center and at the load end of the line are shown in figure 4.13. ■

Example 4.4 We have the same problem as specified in example 4.3, but this time it is assumed that the gates used in figure 4.7 do not allow a current to enter the gate output, as would be the case for ECL. This means that the current variable i cannot be negative.

The procedure for solving this problem is the same as in the previous example. The only difference is that when the gate output-characteristics intersect the vertical axis, they continue in a positive vertical direction along the v-axis. As a consequence, the load-line diagram shown in figure 4.14 differs substantially from that of figure 4.12a. ■

Example 4.5 The left gate shown in figure 4.7 has an output characteristic given by (4.12) and (4.13). The right-hand gate has a load characteristic which can be represented by a resistance of 86 Ω.

The transmission line connecting the two gates has a characteristic impedance $Z_0 = 200\,\Omega$. It is desired to plot the voltage waveforms at the receiving ends of the transmission line for both upward and downward voltage transitions.

The procedure for solving this problem is the same as it has been in the previous examples. To reduce graphical construction, the two transitions have been drawn on one graph, as shown in figure 4.15. From that graph, the voltage waveforms at the load end are readily constructed, and they are shown in figure 4.16. ■

4.4 Series and Parallel Terminated Lines

In figure 4.7 we have the output of one gate driving a transmission line directly. The output of the transmission line is connected directly to

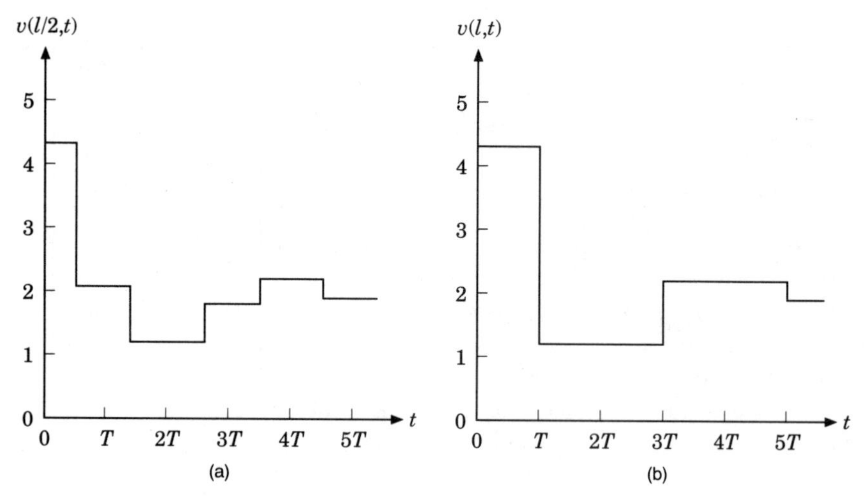

Figure 4.13 Voltage waveforms at the center (a) and receiving end (b) of the line of example 4.3.

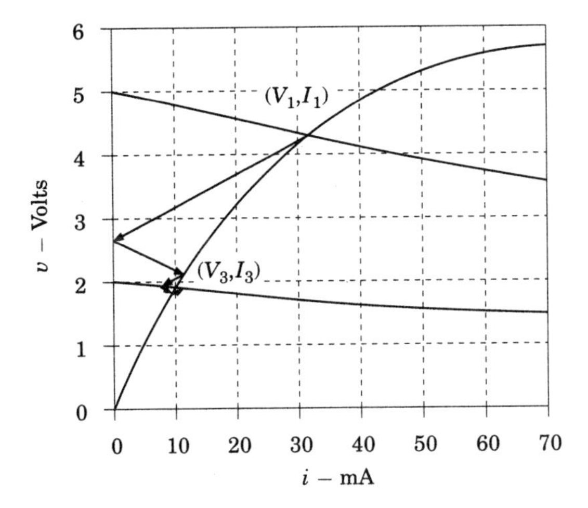

Figure 4.14 Load-line diagram for the problem in example 4.4.

the input of a gate. If a series or shunt termination is used as shown in figure 4.17, then the Bergeron-plot method has to be modified to account for the presence of the terminating resistor.

The modified method is best explained through examples.

Example 4.6 Assume that the left gate shown in figure 4.17a has an output characteristic given by (4.12) and (4.13). The right-hand gate has a load characteristic given by (4.14).

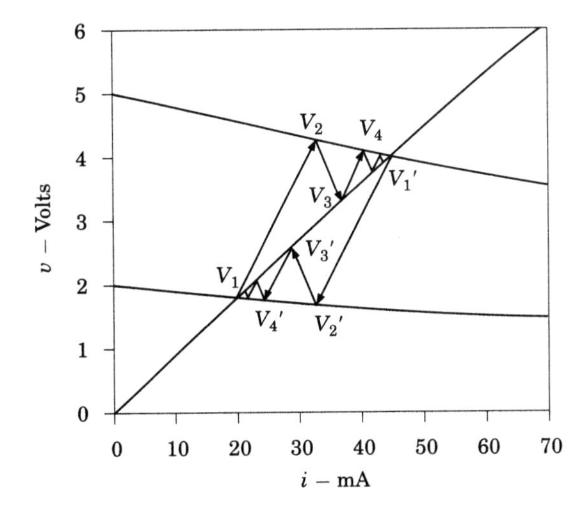

Figure 4.15 Load-line diagram for the problem in example 4.5.

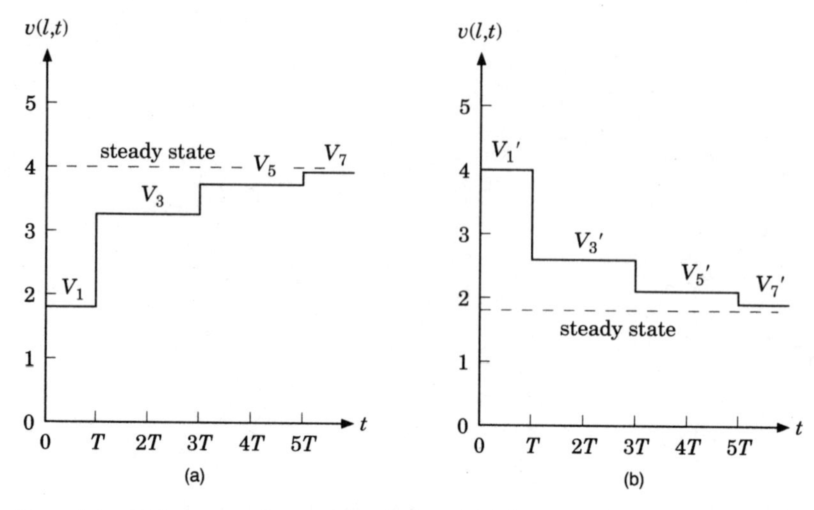

Figure 4.16 Voltage waveforms at the load end (a) for an upward transition and (b) for a downward transition for the problem in example 4.5.

The transmission line connecting the two gates has a characteristic impedance $Z_0 = 50\,\Omega$. It is desired to determine the proper series termination R_s needed, and then to verify the circuit behavior using a Bergeron plot.

In figure 4.10 it is seen that in the vicinity of the high-state characteristic, the Bergeron plot has essentially equal excursions about a current of 40 mA. It is estimated that the high-state characteristic has a slope of $-20\,\Omega$ in that region. From figure 4.12 it can be seen that in the vicinity of the low-state characteristic, the Bergeron plot has essentially equal excursions about a current of 5 mA. It is estimated that the low-state characteristic has a slope of $-10\,\Omega$ in that

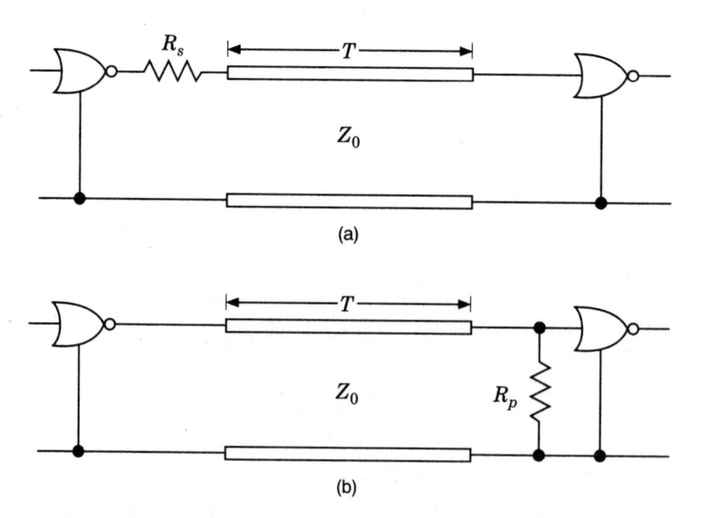

Figure 4.17 Gates connected with a transmission line with (a) series termination and (b) parallel termination.

region. The above two estimates represent an average slope of $-15\,\Omega$. It is decided to augment the $15\,\Omega$ gate output-resistance with the external resistance $R_S = 35\,\Omega$ in an effort to obtain a series match with the $50\,\Omega$ transmission line.

For any current i emerging from the left gate shown in figure 4.17a, the voltage will be lowered by the voltage drop across R_s, which in this case is $35i$ volts. Hence the new parameters for this problem are

$$v = 2e^{-5i} - 35i \qquad \text{in the low state} \tag{4.15}$$

$$v = 5e^{-5i} - 35i \qquad \text{in the high state} \tag{4.16}$$

The load characteristic of the right gate is unchanged, hence it is still given by (4.14).

The voltage-current curves for (4.14), (4.15), and (4.16) are shown plotted in figure 4.18. The Bergeron plot for a positive (upward) logic-transition is shown with the solid lines. The Bergeron plot for a negative (downward) logic-transition is shown with the dashed lines. It is quite clear from this diagram that there is very little oscillation about the steady-state signal at the receiving gate. As is typical of series terminated lines, there is significant signal oscillation at the center of the line as well as in the vicinity of the sending gate. ■

Example 4.7 As in the previous example, the left gate shown in figure 4.17b has an output characteristic given by (4.12) and (4.13). The right-hand gate has an input characteristic given by (4.14).

The transmission line connecting the two gates has a characteristic impedance $Z_0 = 50\,\Omega$. It is desired to determine the proper parallel termination R_p needed, and then to verify the circuit behavior using a Bergeron plot.

In figure 4.10 it is seen that in the region between the two steady-state points the load characteristic of the right gate has an average resistance of approximately $110\,\Omega$. It is decided to place a resistance $R_p = 90\,\Omega$ in shunt

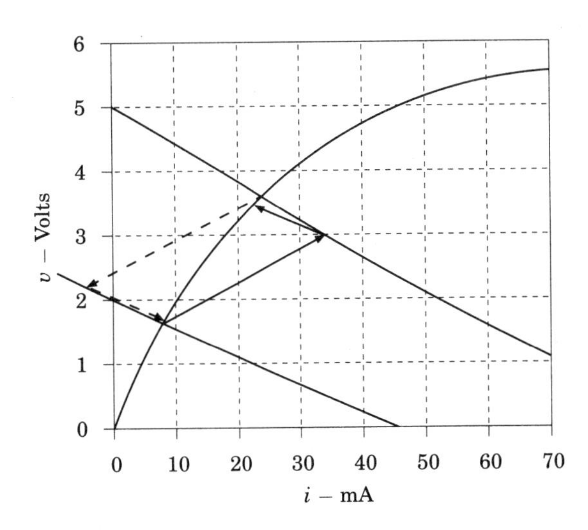

Figure 4.18 Load-line diagram for the problem in example 4.6.

with the input to the receiving gate in an effort to obtain a parallel match with the 50 Ω transmission line.

We can solve (4.14) for the current i which is drawn by the right gate for any voltage v appearing across its input terminals. To this must be added the current $v/90$ to account for the current drawn by the resistance R_p. Hence the equation characterizing the right-hand load is now

$$i = -\frac{1}{40}\ln\left[1 - \frac{v}{6}\right] + \frac{v}{90} \tag{4.17}$$

The output characteristic of the left gate is unchanged, hence it is still given by (4.12) and (4.13).

The voltage-current curves for (4.12), (4.13), and (4.17) are shown plotted in figure 4.19. The Bergeron plots for a positive (upward) logic-transition and for a negative (downward) logic-transition are each shown with a single line. The terminus of each arrow is so close to the steady-state value that there is no point in continuing to clutter up the plot with more arrows. It is quite clear from this diagram that there is very little oscillation about the steady-state signal at the receiving gate. As is typical of shunt terminated lines, what is true for the load is also true for the rest of the transmission line. We see that the method demonstrated in this example is suitable for determining the proper parallel termination. ■

The above two examples were used to demonstrate the use of the Bergeron plot for both series and shunt terminated lines. An examination of the results obtained by the above methods shows that if the gate characteristics are not unusually non-linear in the region of operation, then one could readily switch to a linearized gate-model and still obtain nearly perfect impedance matching for both cases. The next section will demonstrate the application of the methods of the Bergeron-plot method to TTL and to CMOS logic.

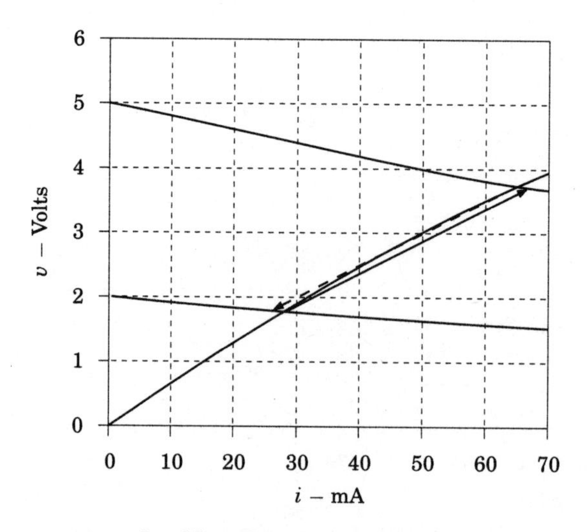

Figure 4.19 Load-line diagram for the problem in example 4.7.

4.5 TTL and CMOS Analysis Using Bergeron Plots

We will now apply the methods presented in the previous sections to solving interface problems encountered when using gates belonging to two very commonly utilized logic families. We will use the input and output voltage-current curves appearing at the end of this chapter. The first set of curves is for the 74F244 chip, a TTL logic buffer/line driver belonging to the Fairchild Advanced Schottky TTL (FAST) family of integrated circuits [4]. The second set is for the 74ACTQ244 chip, a CMOS logic buffer/line driver belonging to the Fairchild Advanced CMOS Technology (FACT) family of chips [5]. (Photocopies of those curves can be used to construct drawings in order to follow the material below.) The reader is cautioned that the current i shown in these curves has the opposite polarity of that used in gate specification-sheets published by National Semiconductor Corporation*.

Both of the chips mentioned above have the same input thresholds. A voltage below $V_{ILmax} = 0.8\,\mathrm{V}$ is reliably recognized as a low-state input-voltage. A voltage above $V_{IHmin} = 2.0\,\mathrm{V}$ is reliably recognized as a high-state input-voltage. The range between the above two voltages may be regarded as a fuzzy region for which the state of the chips cannot be reliably determined. As has been done previously, examples will be used to illustrate the applications.

Example 4.8 Assume that the gates shown in figure 4.7 are both of the 74F244 buffer/driver variety whose characteristic curves can be found at the end of the chapter. The two gates are connected with a $50\,\Omega$ transmission line. It is desired to plot the voltage waveforms at the receiving-gate end of the transmission line.

The solution to this problem is found in figure 4.20. The Bergeron plot for a negative (downward) logic-transition is shown with the dashed lines. It is quite clear that the low-voltage steady-state of $0.2\,\mathrm{V}$ is reached after the reflection activity indicated in the diagram. The Bergeron plot for a positive (upward) logic-transition is shown with the solid lines. Here a minor problem arises due to the fact that the high-state output-characteristic and the input characteristic lie so close to one another near the vertical axis in the region $3.7\,\mathrm{V} \leq v \leq 6\,\mathrm{V}$. When the Bergeron plot reaches the vicinity of $5\,\mathrm{V}$ it is seen that many reflections are then needed to make the diagram reach the final steady-state of $3.7\,\mathrm{V}$. Hence the final steady-state voltage may not be reached for a long time.

Figure 4.21 is a direct result of the Bergeron plots shown in figure 4.20. It shows the voltage waveforms at the receiving gate for both positive and negative voltage-transitions. Although there is ringing, the signal for the upward transition is at no time below the minimum high-threshold input-voltage $V_{IHmin} = 2.0\,\mathrm{V}$. Similarly during a downward transition, the input voltage is never above the maximum low-threshold input-voltage $V_{ILmax} = 0.8\,\mathrm{V}$. So there

*Fairchild Corporation products and specifications are now available through National Semiconductor Corporation, Santa Clara, California.

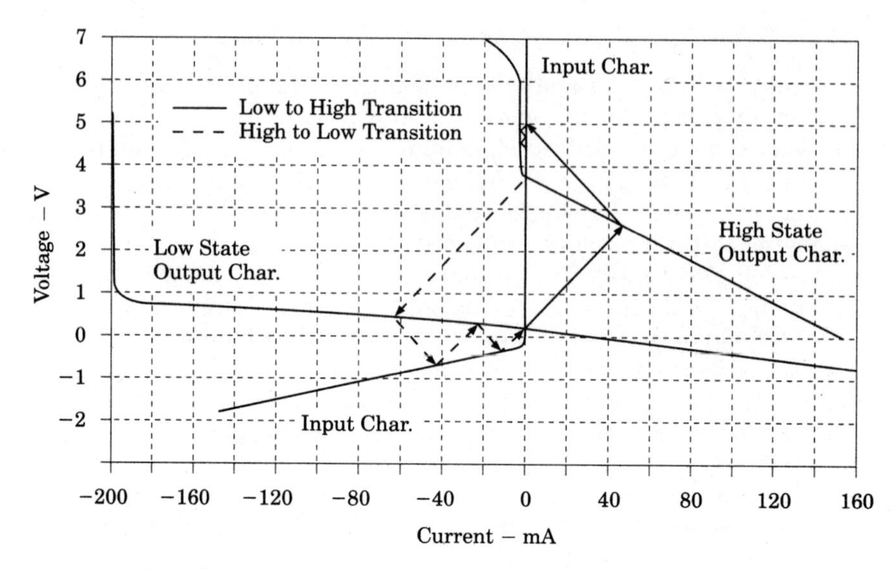

Figure 4.20 Load-line diagram for the 74F244 buffer/driver chip used with a 50 Ω transmission line.

is no question that the receiving gate will respond properly to the digital signals it receives, even with an unterminated line. If the oscillations are objectionable for some reason, then the line can be series terminated as illustrated in the next example. ■

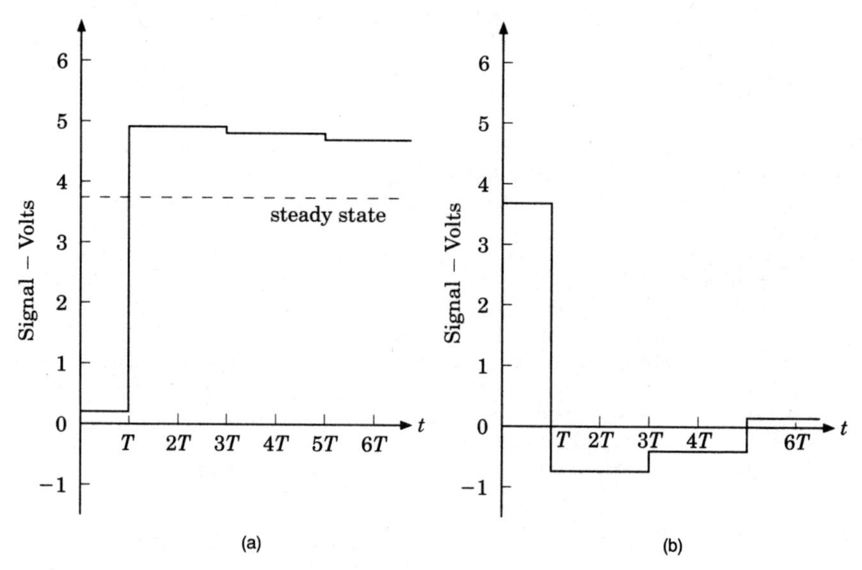

Figure 4.21 Voltage waveform at the receiving end of the transmission line for (a) a positive and (b) a negative voltage transition for example 4.8.

Example 4.9 This is the same problem as in the previous example, except that it is desired to add a series termination to get rid of the oscillations that appear in figure 4.21. Since the output impedance of the low-state output-characteristic differs appreciably from that of the high-state output-characteristic, we will attempt to create an impedance match only for the latter. It is hoped that the resultant high-to-low transition will be well behaved anyway. The high-state output-characteristic exhibits an output impedance of $25\,\Omega$. To match the $50\,\Omega$ characteristic impedance of the line, we choose $R_S = 25\,\Omega$.

The new output characteristics are shown in figure 4.22 along with the Bergeron plot. We see that the high steady-state is reached after one reflection, as is the case for series matched lines. The circuit undershoots the low steady-state slightly and requires an additional, quite insignificant, reflection to get to its final value. ■

Example 4.10 This time we will assume that the gates shown in figure 4.7 are both of the 74ACTQ244 buffer/driver type. The two gates are connected with a $50\,\Omega$ transmission line. It is desired to plot the voltage waveforms at the receiving-gate end of the transmission line.

The solution to this problem appears in figure 4.23. The Bergeron plot for this case is very straightforward and is not complicated by minor details as was the case in example 4.8.

Figure 4.24 is a direct result of the Bergeron plots shown in figure 4.23. It shows the voltage waveforms at the receiving gate for both positive and negative voltage-transitions. The waveform shows ringing, but the received signal for the upward transition is at no time below the minimum high-threshold input-voltage $V_{IHmin} = 2.0\,\text{V}$. Similarly during a downward transition, the input voltage is never above the maximum low-threshold input-voltage $V_{ILmax} = 0.8\,\text{V}$. So there

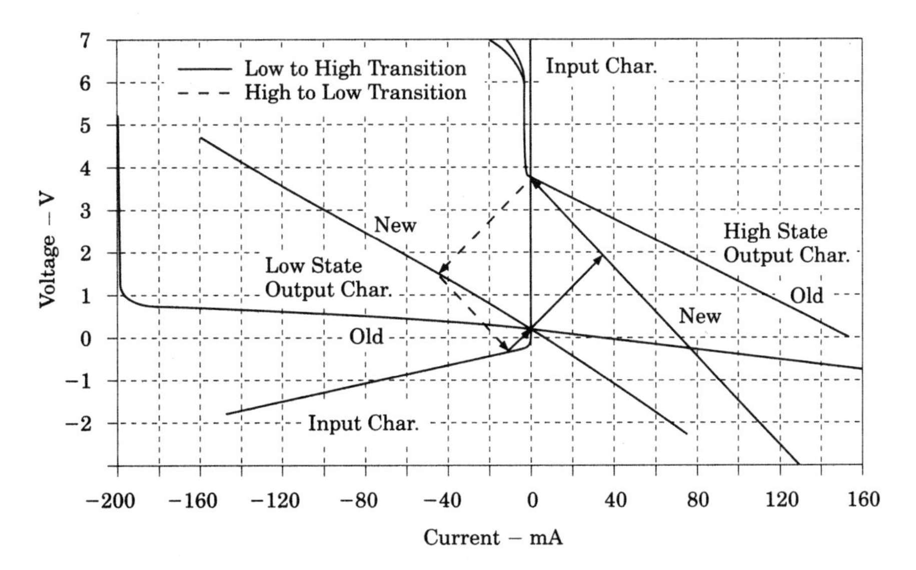

Figure 4.22 Bergeron plot for example 4.9.

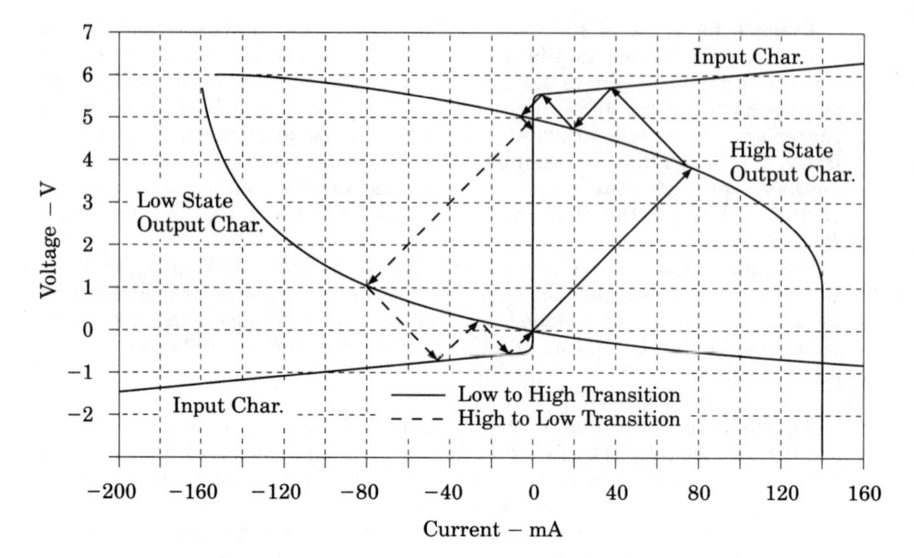

Figure 4.23 Load-line diagram for the 74ACTQ244 buffer/driver chip used with a 50 Ω transmission line.

is no doubt that the receiving gate will respond correctly to the digital signals it receives, even with an unterminated line. In the next example an attempt is made to get rid of the oscillations by series terminating the transmission line. ■

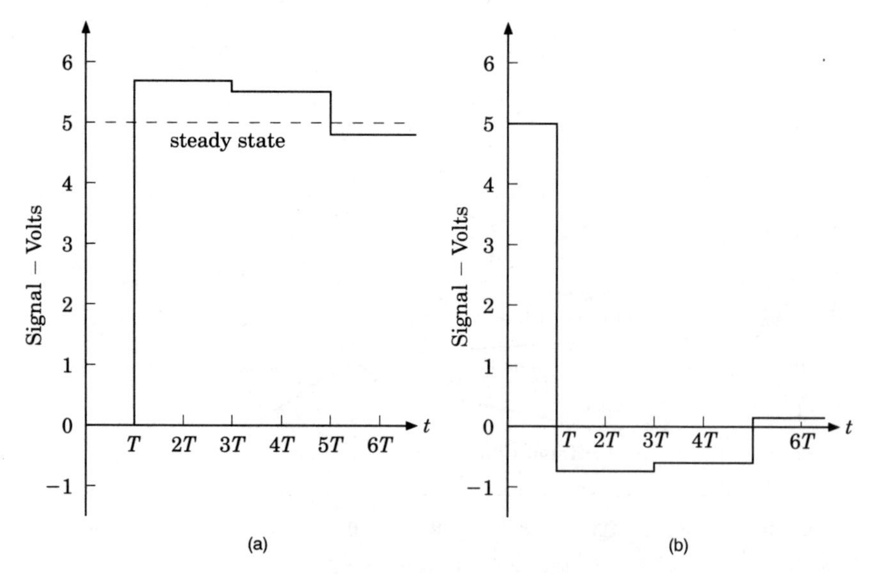

Figure 4.24 Voltage waveform at the receiving end of the transmission line for (a) a positive and (b) a negative voltage transition for example 4.10.

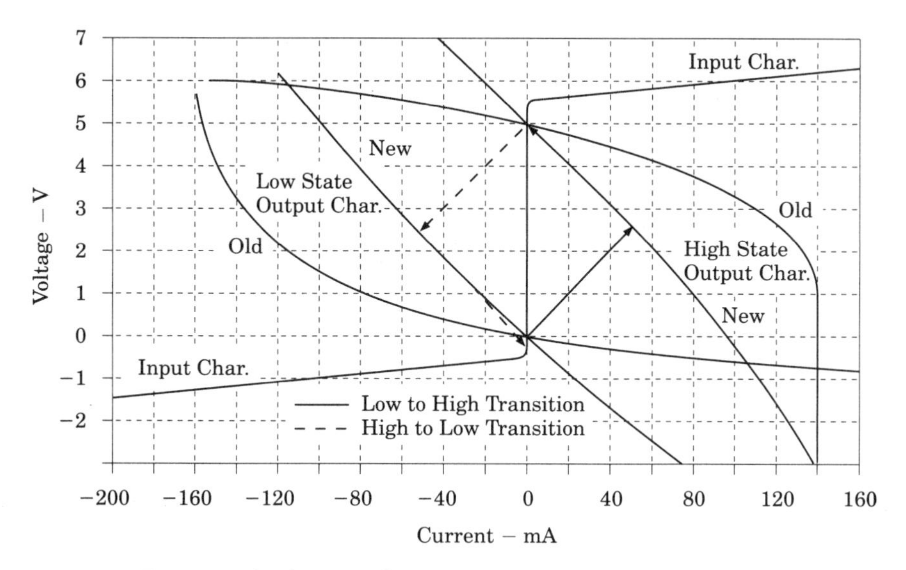

Figure 4.25 Bergeron plot for example 4.11.

Example 4.11 This is the same problem as in the previous example, except that it is desired to add a series termination to get rid of the oscillations that appear in figure 4.24. In figure 4.23 we find that where the Bergeron plot exhibits its most significant behavior both output characteristics have an output impedance of approximately $15\,\Omega$. To match the $50\,\Omega$ characteristic impedance of the line, we choose $R_S = 35\,\Omega$.

The new output characteristics are shown in figure 4.25 along with the Bergeron plot. We see that both steady states are reached after one reflection, as can be expected for series matched lines. ■

4.6 Conclusion

This chapter was used to present the method of load-line analysis also known as the Bergeron-plot method. Familiarity of this method means that we are no longer constrained to solving transmission-line problems for linear sources and for linear loads. Now we are in a position to analyze problems for which the methods of chapter 2 fail. We can now deal with problems in which the output characteristics as well as the input characteristics of the devices connected to the transmission lines are non-linear. The Bergeron-plot method thus becomes another tool for analysis as well as a tool for developing insights into otherwise difficult transmission-line problems.

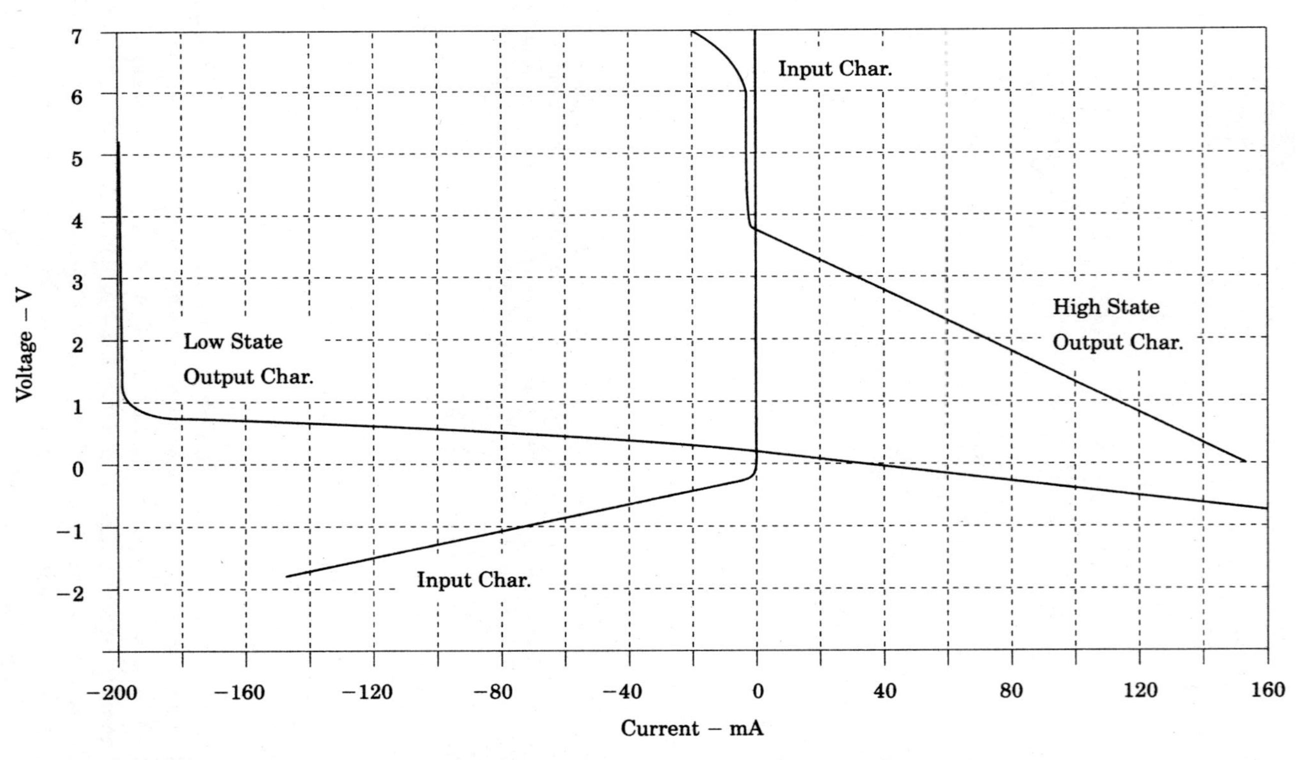

Figure 4.26 Input and output characteristics for the 74F244 chip, a TTL logic buffer/line driver.

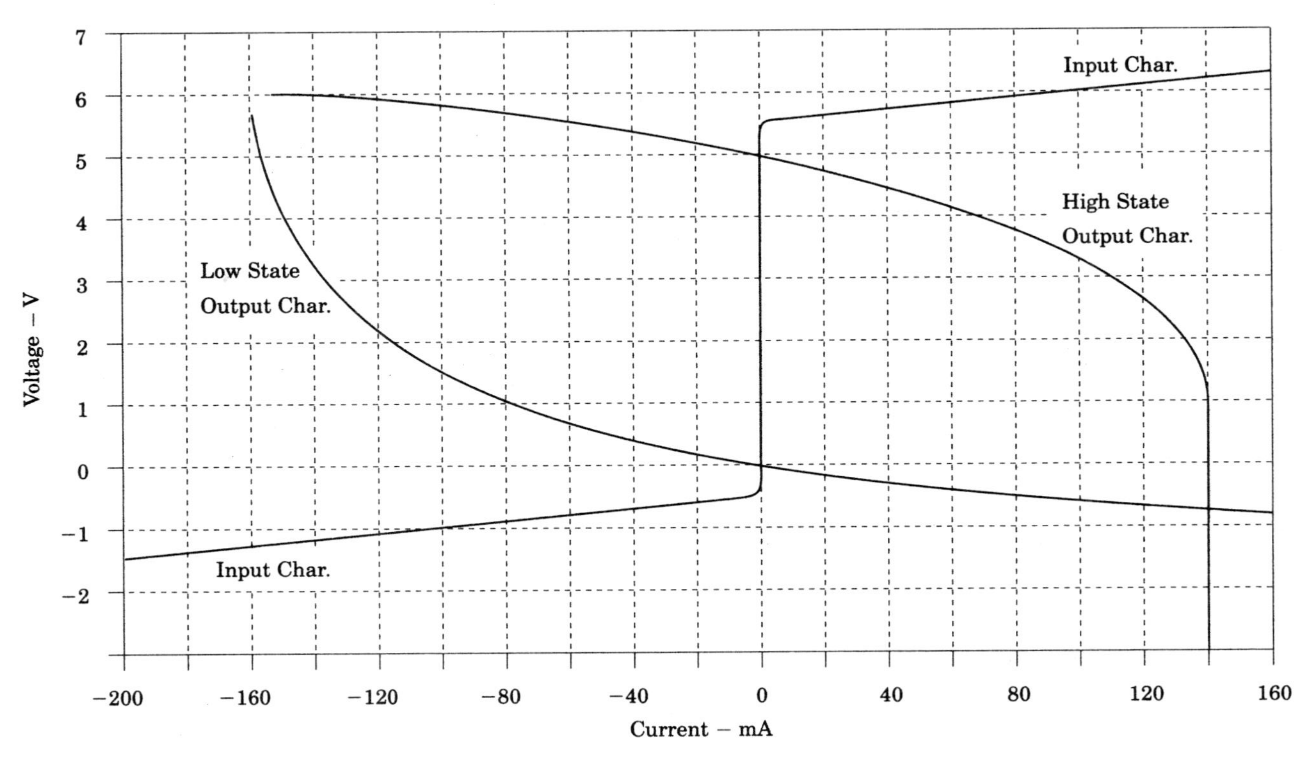

Figure 4.27 Input and output characteristics for the 74ACTQ244 chip, a CMOS logic buffer/line driver.

References

1. *Transmission Line Effects in PCB Applications,* Motorola Semiconductor Application Note AN1051, Motorola Literature Distribution, P.O. Box 20912, Phoenix, Arizona 85036, 1990.
2. Robert A. Stehlin, "Bergeron Plots Predict Delays in High-Speed TTL Circuits," *EDN,* November 15, 1984, pp. 293–298.
3. Robert S. Singleton, "No Need to Juggle Equations to find Reflection — Just Draw Three Lines," *Electronics,* October 28, 1968, pp. 93–99.
4. *FAST Logic Applications Handbook,* 1990 Edition, National Semiconductor Corporation, Santa Clara, California.
5. *FACT Advanced CMOS Logic Databook,* 1990 Edition, National Semiconductor Corporation, Santa Clara, California.

Problems

P4.1 We have a gate whose output can be modeled as a 5 V source with an output impedance of 1 Ω which is driving a 3 Ω load. The equivalent circuit of this connection is shown in figure 4.3a.

(a) Solve analytically for the voltage and current at the load.
(b) Obtain the same solution using the load-line (Bergeron-plot) analysis.

P4.2 We have a gate whose voltage-current output characteristic is given by the relationship $v = 12\cos(\pi i/0.06)$. This is connected to a load whose voltage-current characteristic is $v = 5\log_{10}[1000(i + 0.001)]$. The equivalent circuit of this connection is shown in figure 4.4a.

(a) Solve analytically for the voltage and current at the load.
(b) Obtain the solution using load-line analysis.

P4.3 Example 4.4 can be used to get the practice of obtaining the lattice diagram from the Bergeron plot.

(a) Plot the voltage lattice diagram for this example.
(b) Using the lattice diagram, plot the voltage waveforms at the source and load ends of the transmission line.

P4.4 To get the practice in the use of the Bergeron-plot method, redo example 4.2 for a 25 Ω transmission line.

P4.5 To get more practice in the use of the Bergeron-plot method, redo example 4.3 for a 25 Ω transmission line.

P4.6 In connection with example 4.6, plot the voltage at the sending end, the center, and the receiving end of the line.

P4.7 To get the practice in the use of the modified Bergeron-plot method, redo example 4.6 for a 25 Ω transmission line.

P4.8 To get more practice in the use of the modified Bergeron-plot method, redo example 4.7 for a 25 Ω transmission line.

P4.9 Redo the TTL problem of example 4.8 for a transmission line with $Z_0 = 25\,\Omega$.

P4.10 In connection with the Bergeron plot figure 4.22, plot the voltage at the sending end and at the center of the line.

P4.11 In example 4.9 a series match was presented for example 4.8. Make an attempt at parallel matching for the same problem.

P4.12 Redo the CMOS problem of example 4.10 for a transmission line with $Z_0 = 25\,\Omega$.

P4.13 In example 4.11 a series match was presented for example 4.10. Make an attempt at parallel matching for the same problem.

Chapter

5

Crosstalk on Transmission Lines

5.1 Introduction

Crosstalk is the term used for signals coupled from one transmission line into another by time varying signals. The coupling, both capacitive and inductive, is generally due to the proximity of the two lines. This coupling is so undesirable that the strong terms *aggressor line* and *victim line* are used to describe the two interacting transmission-lines. This problem occurs on printed-circuit boards as well as on twisted-pair cables. An example of such coupling appears in the printed circuit board (PCB) representation of the two closely-spaced transmission-lines shown in figure 5.1.

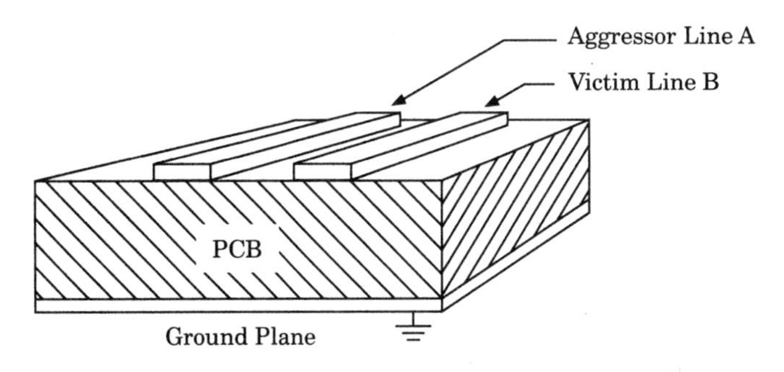

Figure 5.1 Two closely spaced PCB transmission lines.

5.2 Derivation of the Basic Crosstalk Relations

To simplify the analysis of this problem we will use the model shown in figure 5.2. Here we have shown two transmission lines of characteristic impedance Z_0 in close proximity. The two lines have a common ground. The top line, which is excited by a source, is the aggressor line. The bottom line, which is passive, is the victim line. The ungrounded conductors of both lines are sufficiently close to have mutual capacitive (electric) coupling of C_M farads/meter and mutual inductive (magnetic) coupling of L_M henries/meter. To keep the analysis simple, both lines are terminated in their characteristic impedance, so that any signals propagating in either direction on the lines cause no reflections. It is assumed that the coupling between the two lines is weak, so we will only concern ourselves with the crosstalk from line A induced in line B, but not with the crosstalk into line A due to the signals induced in line B.

As in all previous chapters, distance along the transmission line is expressed in terms of the variable z. The point on the line where crosstalk is taking place will be designated by the variable ζ. We will think of a signal propagating on aggressor line A arriving at the specific point $z = \zeta$ where it makes a differential contribution to the crosstalk signal on victim line B. The differential segment of line at $z = \zeta$ whose inductance and capacitance are responsible for the crosstalk has a length $\Delta\zeta$. To obtain the entire crosstalk signal it is necessary to sum the differential contributions for all values of ζ.

We will first derive the basic relation pertaining to capacitive crosstalk. We start the analysis by considering a section of both transmission lines, of width $\Delta\zeta$, which is located at the specific location $z = \zeta$, as shown in figure 5.3. The two lines are shown joined by a differential capacitance of value $C_M\Delta\zeta$. Transmission line B is assumed to be uncharged. A differential crosstalk current flows into

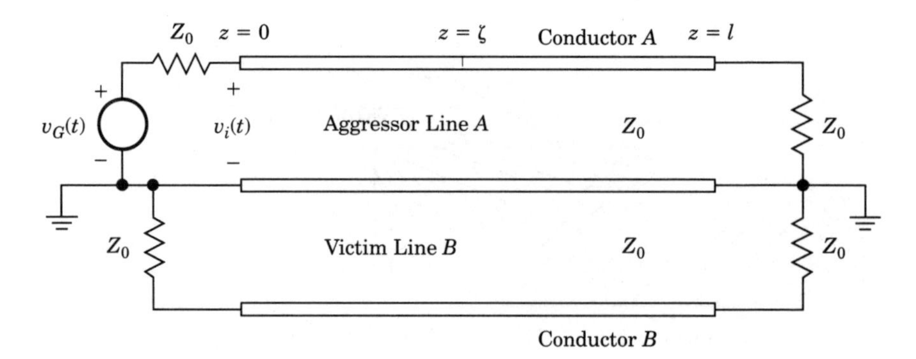

Figure 5.2 Model of two lines in close proximity.

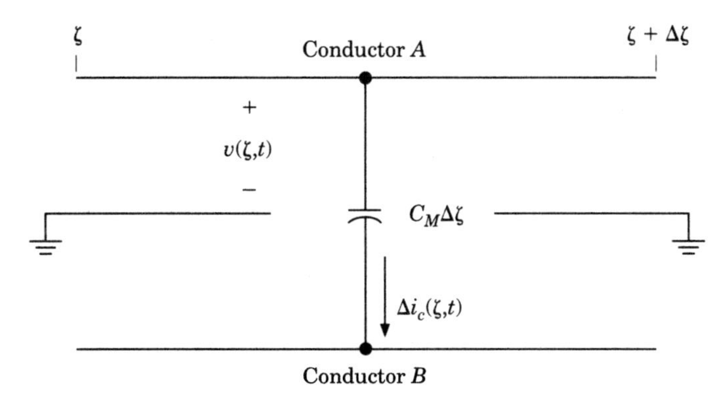

Figure 5.3 Schematic of a differential section of capacitively-coupled transmission-lines.

conductor B as a consequence of the voltage difference between the two conductors. Its value is

$$\Delta i_c(\zeta,t) = C_M \Delta\zeta \frac{\partial v(\zeta,t)}{\partial t} \tag{5.1}$$

This differential crosstalk-current is modeled by an ideal current source connected from the lower to the upper conductor of transmission line B, as shown in figure 5.4. Strictly speaking, the negative end of the current source should be shown attached to the upper conductor of transmission line A to be in agreement with figure 5.3. But the solution to the problem is not at all affected by showing the negative end of the current source attached to the ground conductor of transmission line B. The differential crosstalk-current at ζ sees the characteristic impedance Z_0 to *both* the left and right of ζ, hence it splits into two parts, with one half propagating to the right, and the other half propagating to the left. The two current waves, of value $\Delta i_c(\zeta,t)/2$, are accompanied by two voltage waves. The two voltage waves, which are of like polarity, are shown in figure 5.4 propagating away from the point $z = \zeta$. They are calculated by taking

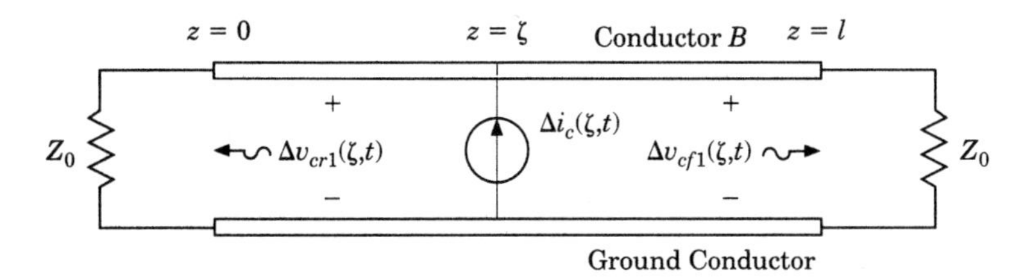

Figure 5.4 Differential crosstalk current due to capacitive coupling shown feeding into victim line B.

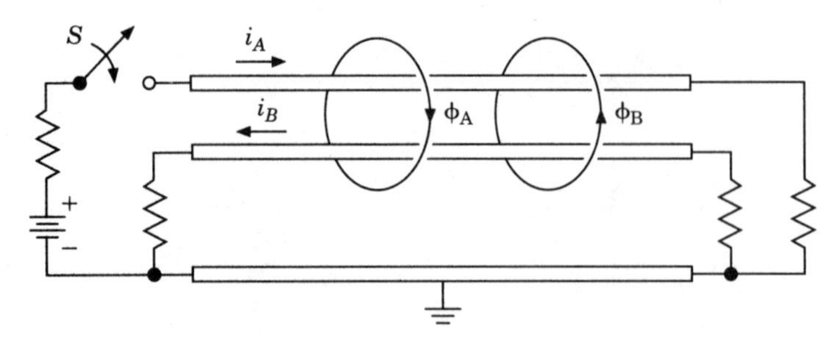

Figure 5.5 Model for reviewing Lenz's law.

(5.1), dividing by 2, and multiplying by Z_0. At $z = \zeta$ their values are

$$\Delta v_{cr1}(\zeta, t) = \Delta v_{cf1}(\zeta, t) = \frac{C_M \Delta\zeta}{2} \frac{\partial v(\zeta, t)}{\partial t} Z_0 \tag{5.2}$$

Before we proceed with the analysis of inductively coupled crosstalk, we will first review Lenz's law.

In figure 5.5 are shown two transmission lines in close proximity. When switch S is closed, the direct-voltage source on the left side tries to establish the current i_A in the top conductor. This current is accompanied by a rising magnetic flux ϕ_A. Lenz's law states that the lower conductor will respond in such a way as to counteract the changes caused by the buildup of current in the top conductor. The lower conductor will therefore have a current induced in it with a polarity appropriate for the magnetic flux ϕ_B, which will attempt to cancel out the magnetic flux ϕ_A. The current i_B, and its polarity, are shown in figure 5.5. With this thumbnail review out of the way we can now return to the problem of inductively induced crosstalk.

As we did for the capacitively coupled analysis, we start by considering a section of both transmission lines, of width $\Delta\zeta$, which is located at $z = \zeta$, as shown in figure 5.6.

The two lines shown are magnetically coupled with a mutual inductance of value $L_M \Delta\zeta$. The dot orientation shown is in accordance with Lenz's law and is justified as follows. A rising current in line A will cause the upper inductor to assume a positive voltage polarity at the dot. This will cause the induced voltage in conductor B also to have positive polarity at the dot. The resultant current flow in line B will then try to cancel the magnetic field that conductor A is trying to establish.

Transmission line B is assumed to be deenergized, hence there is no previous voltage or current on line B. A differential crosstalk-voltage is induced in line B with the polarity shown. Its value is

$$\Delta v_c(\zeta, t) = L_M \Delta\zeta \frac{\partial i(\zeta, t)}{\partial t} \tag{5.3}$$

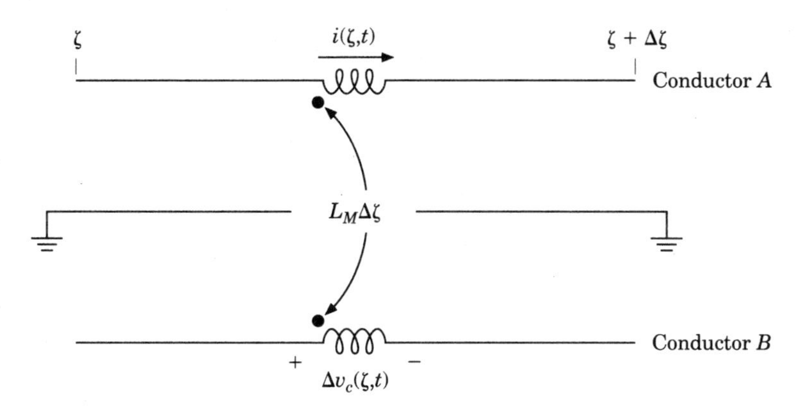

Figure 5.6 Schematic of a differential section of inductively-coupled transmission-lines.

This differential crosstalk-voltage is modeled by a voltage source as shown in figure 5.7. The differential crosstalk-voltage at ζ is connected at each end to a transmission line of characteristic impedance Z_0. Hence it encounters a total impedance of $2Z_0$. The current that flows at ζ is therefore the voltage of (5.3) divided by $2Z_0$. This local current creates the two current waves, as shown in figure 5.7, propagating away from $z = \zeta$. At $z = \zeta$ their values are

$$\Delta i_{cr2}(\zeta,t) = -\Delta i_{cf2}(\zeta,t) = \frac{1}{2Z_0} L_M \Delta\zeta \frac{\partial i(\zeta,t)}{\partial t} \tag{5.4}$$

Note that $\Delta i_{cf2}(\zeta,t)$ is in fact negative since its direction of flow is contrary to that indicated in figure 5.7.

Since propagating voltage and current waves on transmission lines are related by the characteristic impedance Z_0, we find the values of the two voltage waves by simply multiplying (5.4) by the characteristic

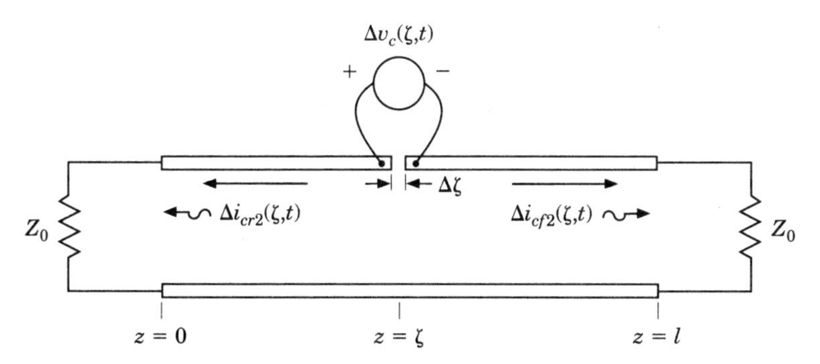

Figure 5.7 Differential crosstalk voltage due to inductive coupling shown exciting victim-line *B*.

impedance of transmission line B, which in our model is equal to Z_0, to obtain

$$\Delta v_{cr2}(\zeta,t) = -\Delta v_{cf2}(\zeta,t) = \frac{1}{2} L_M \Delta\zeta \frac{\partial i(\zeta,t)}{\partial t} \tag{5.5}$$

But on transmission line A, voltage and current waves are also related by the characteristic impedance, which in our model is Z_0. This allows us to replace $i(\zeta,t)$ with $v(\zeta,t)/Z_0$, for the final result

$$\Delta v_{cr2}(\zeta,t) = -\Delta v_{cf2}(\zeta,t) = \frac{1}{2Z_0} L_M \Delta\zeta \frac{\partial v(\zeta,t)}{\partial t} \tag{5.6}$$

Summing the capacitively and inductively induced voltages from (5.2) and (5.6), we get for the forward and reverse crosstalk voltages at some point on the line $z = \zeta$

$$\Delta v_{cf}(\zeta,t) = \frac{1}{2}\left(C_M Z_0 - \frac{L_M}{Z_0}\right)\frac{\partial v(\zeta,t)}{\partial t}\Delta\zeta \tag{5.7}$$

$$\Delta v_{cr}(\zeta,t) = \frac{1}{2}\left(C_M Z_0 + \frac{L_M}{Z_0}\right)\frac{\partial v(\zeta,t)}{\partial t}\Delta\zeta \tag{5.8}$$

We define the forward and reverse crosstalk coefficients as

$$K_{cf} \equiv \frac{1}{2}\left(C_M Z_0 - \frac{L_M}{Z_0}\right) \qquad \text{(seconds/meter)} \tag{5.9}$$

$$K_{cr} \equiv \frac{\nu}{4}\left(C_M Z_0 + \frac{L_M}{Z_0}\right) \qquad \text{(dimensionless)} \tag{5.10}$$

and we observe that K_{cf} can be positive, negative, and even zero. The ν appearing in the last equation represents the speed of signal propagation on the transmission lines. K_{cr} is defined in a seemingly peculiar manner, but it will become clear later that this is done in order to simplify the final crosstalk equations.

Using the above definitions and also the notation

$$v'(\zeta,t) \equiv \frac{\partial v(\zeta,t)}{\partial t} \tag{5.11}$$

in (5.7) and (5.8) we rewrite the two *basic crosstalk relations* in the more succinct form

$$\Delta v_{cf}(\zeta,t) = K_{cf} v'(\zeta,t)\Delta\zeta \tag{5.12}$$

$$\Delta v_{cr}(\zeta,t) = \frac{2}{\nu} K_{cr} v'(\zeta,t)\Delta\zeta \tag{5.13}$$

The above are the basic crosstalk equations, and are by no means the final solutions to the crosstalk problem. The derivation of the forward crosstalk equation will be undertaken in the next section and that for the reverse crosstalk will be taken up in the section that follows.

5.3 The Forward Crosstalk Equation

Assume that we have an incident wave $v_i(t - z/\nu)$ traveling from left to right on aggressor line A, as is shown in figure 5.8. On transmission line A this wave produces at $z = \zeta$ a time waveform described by $v_i(t - \zeta/\nu)$. According to (5.12), we have a forward crosstalk waveform on victim line B, at ζ, given by

$$\Delta v_{cf}(\zeta, t) = K_{cf} v_i'\left(t - \frac{\zeta}{\nu}\right)\Delta\zeta \tag{5.14}$$

When this waveform reaches an observer standing on line B, at a point z which is to the right of ζ, it is delayed by $(z - \zeta)/\nu$ seconds. Hence the argument in (5.14) has to be corrected by this amount. We get

$$\Delta v_{cf}(z, t) = K_{cf} v_i'\left(t - \frac{\zeta}{\nu} - \frac{z - \zeta}{\nu}\right)\Delta\zeta \tag{5.15}$$

Again it is reiterated that the term $(z - \zeta)/\nu$ represents the additional delay required for the signal to propagate from its point of origination ζ, to the observer's location at z. The above simplifies to

$$\Delta v_{cf}(z, t) = K_{cf} v_i'\left(t - \frac{z}{\nu}\right)\Delta\zeta \tag{5.16}$$

which in the limit as $\Delta\zeta$ goes to zero produces the integral

$$v_{cf}(z, t) = K_{cf} \int_0^z v_i'\left(t - \frac{z}{\nu}\right)d\zeta \tag{5.17}$$

Although aggressor line A can extend to the left of $z = 0$, this is the leftmost coordinate on victim line B. Hence the mutual coupling

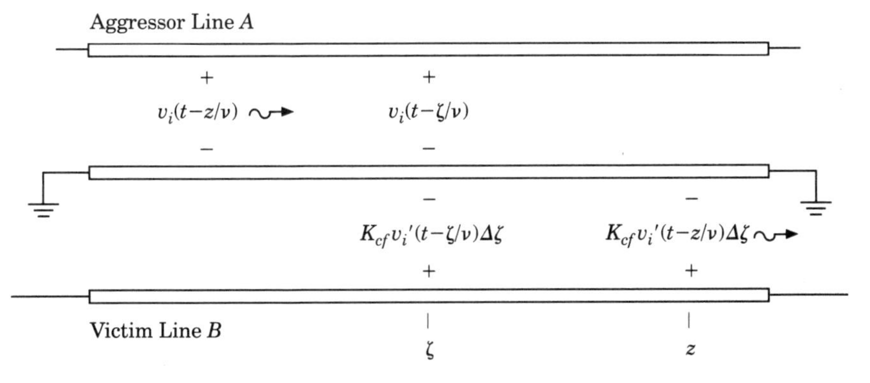

Figure 5.8 The creation of the forward crosstalk voltage.

between lines A and B begins at $z = 0$ which is the justification for the lower limit of integration. The upper limit of integration corresponds to the point z at which the observer is located. The waveform $v_i(\cdot)$ on line A makes no further contribution to the forward crosstalk after it has passed the observer who is located at point z. Once the waveform $v_i(\cdot)$ passes the observer at z it still continues to make crosstalk contributions on line B which propagate to the right, away from the observer, hence they cannot be sensed by the observer.

The integrand in (5.17) is not a function of the variable of integration ζ, hence it can be taken outside the integral. The final result of the integration is the forward-crosstalk equation

$$v_{cf}(z,t) = z\,K_{cf}\,v_i'\left(t - \frac{z}{\nu}\right) \tag{5.18}$$

where, as mentioned in (5.11), the prime in the expression $v_i'(\cdot)$ denotes differentiation with respect to time.

Example 5.1 We have the same arrangement as shown in figure 5.2. The source $v_G(t)$ has the appropriate waveform to create $v_i(t)$, the voltage at the input to the transmission line shown in figure 5.9a, and its derivative is shown in figure 5.9b. It is assumed (to simplify the drawings) that the signal transition time a is smaller than the transmission-line transit-time T and that K_{cf} is positive. We wish to plot the forward crosstalk waveforms at $z = \frac{1}{4}l$, $z = \frac{1}{2}l$, and $z = \frac{3}{4}l$.

To solve this problem we simply evaluate (5.18) for the three desired values of z to obtain

$$v_{cf}\left(\frac{1}{4}\,l,t\right) = \frac{1}{4}\,lK_{cf}v_i'\left(t - \frac{1}{4}\,T\right) \tag{5.19}$$

$$v_{cf}\left(\frac{1}{2}\,l,t\right) = \frac{1}{2}\,lK_{cf}v_i'\left(t - \frac{1}{2}\,T\right) \tag{5.20}$$

$$v_{cf}\left(\frac{3}{4}\,l,t\right) = \frac{3}{4}\,lK_{cf}v_i'\left(t - \frac{3}{4}\,T\right) \tag{5.21}$$

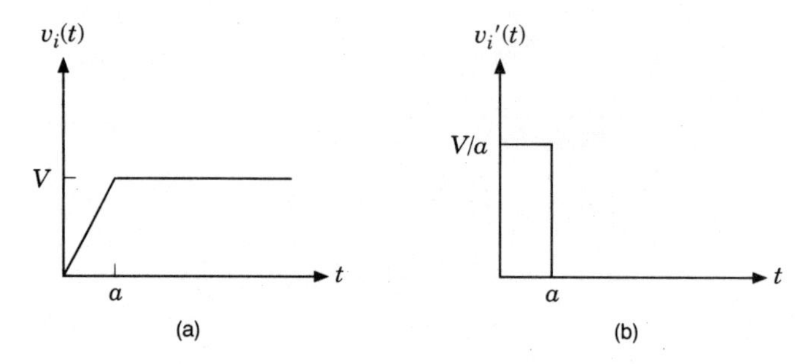

Figure 5.9 (a) The input voltage on aggressor line A (b) and its time derivative.

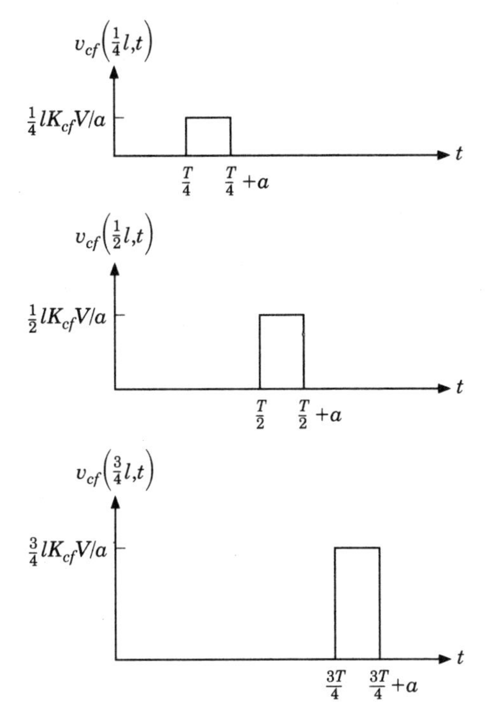

Figure 5.10 The forward crosstalk waveforms at $z = \frac{1}{4}\,l$, $z = \frac{1}{2}\,l$, and $z = \frac{3}{4}\,l$.

All that is left to be done is to use the waveform $v_i'(t)$ (shown in figure 5.9b) in the above equations to obtain the solutions shown in figure 5.10. ■

5.4 The Reverse Crosstalk Equation

Assume that we have an incident wave $v_i(t - z/\nu)$ traveling from left to right on aggressor line A, as is shown in figure 5.11. On transmission line A this wave produces a time waveform at ζ described by $v_i(t - \zeta/\nu)$. According to (5.13), we have a reverse crosstalk waveform on victim line B, at $z = \zeta$, given by

$$\Delta v_{cr}(\zeta, t) = \frac{2}{\nu} K_{cr} v_i'\left(t - \frac{\zeta}{\nu}\right)\Delta\zeta \tag{5.22}$$

The reverse crosstalk waveform propagates in the direction of decreasing z. When this waveform reaches an observer standing on line B, at a point z which is to the left of ζ, it takes on the form

$$\Delta v_{cr}(z, t) = \frac{2}{\nu} K_{cr} v_i'\left(t - \frac{\zeta}{\nu} - \frac{\zeta - z}{\nu}\right)\Delta\zeta \tag{5.23}$$

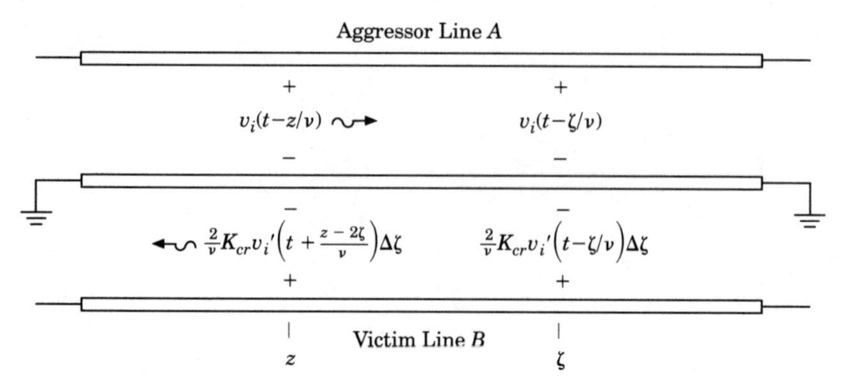

Figure 5.11 The creation of the reverse crosstalk voltage.

The term $(\zeta - z)/\nu$, which is positive since z is to the left of ζ, represents the additional delay required for the signal to propagate from ζ to the observer's location at z. The above reduces to

$$\Delta v_{cr}(z,t) = \frac{2}{\nu}K_{cr}v_i'\left(t + \frac{z - 2\zeta}{\nu}\right)\Delta\zeta \tag{5.24}$$

which in the limit as $\Delta\zeta$ goes to zero produces the integral

$$v_{cr}(z,t) = \frac{2}{\nu}K_{cr}\int_z^l v_i'\left(t + \frac{z - 2\zeta}{\nu}\right)d\zeta \tag{5.25}$$

The lower limit of integration corresponds to the point z at which the observer is located. The waveform $v_i(\cdot)$ on line A makes no contribution to the reverse crosstalk at point z until it has reached that point. The signal $v_i(\cdot)$ continues to make contributions to the reverse crosstalk waveform until it finally goes out of existence when it reaches the end of the transmission line at $z = l$. The result of the integration is

$$v_{cr}(z,t) = -K_{cr}v_i\left(t + \frac{z - 2\zeta}{\nu}\right)\Bigg|_z^l \tag{5.26}$$

which, when evaluated at the limits becomes

$$v_{cr}(z,t) = K_{cr}\left[v_i\left(t - \frac{z}{\nu}\right) - v_i\left(t - 2T + \frac{z}{\nu}\right)\right] \tag{5.27}$$

Example 5.2 We have the same problem as we considered in example 5.1, but this time we are interested in finding the *reverse crosstalk* waveforms at $z = \frac{1}{4}l$, $z = \frac{1}{2}l$, and $z = \frac{3}{4}l$.

Using (5.27) we readily obtain the three expressions

$$v_{cr}\left(\frac{1}{4}\,l, t\right) = K_{cr}\left[v_i\left(t - \frac{1}{4}T\right) - v_i\left(t - \frac{7}{4}T\right)\right] \tag{5.28}$$

$$v_{cr}\left(\frac{1}{2}\,l, t\right) = K_{cr}\left[v_i\left(t - \frac{1}{2}T\right) - v_i\left(t - \frac{3}{2}T\right)\right] \tag{5.29}$$

$$v_{cr}\left(\frac{3}{4}\,l, t\right) = K_{cr}\left[v_i\left(t - \frac{3}{4}T\right) - v_i\left(t - \frac{5}{4}T\right)\right] \tag{5.30}$$

These three waveforms are shown in figure 5.12. ■

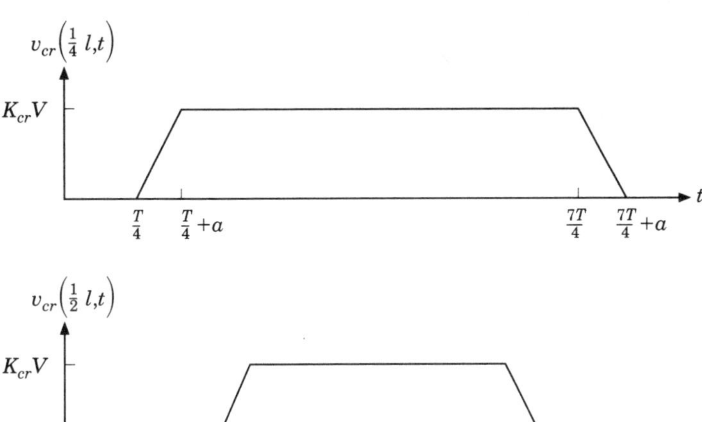

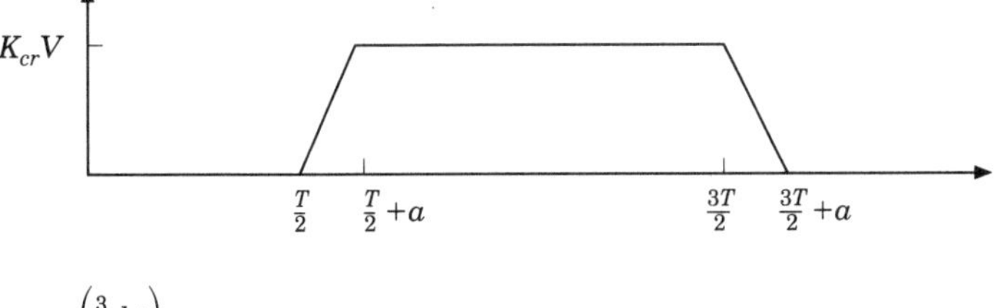

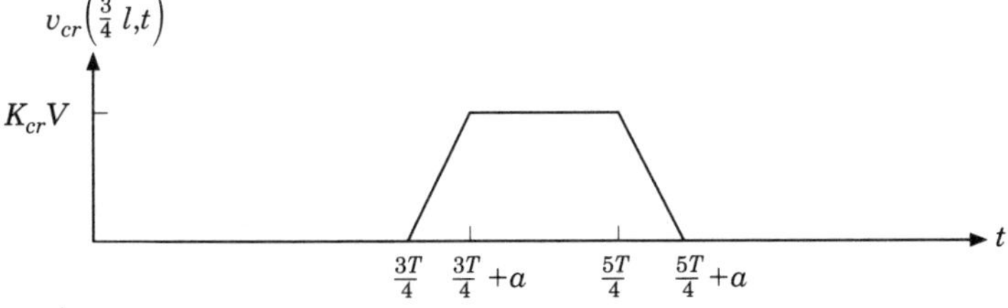

Figure 5.12 The reverse crosstalk waveforms at $z = \frac{1}{4}\,l$, $z = \frac{1}{2}\,l$, and $z = \frac{3}{4}\,l$.

5.5 Unmatched Aggressor Lines

The forward crosstalk equation (5.18) and the reverse crosstalk equation (5.27) were derived for the case of matched aggressor lines. The solutions can be readily extended to cover unmatched aggressor lines. We will assume that in figure (5.2) the resistors of value Z_0 terminating aggressor line A are replaced with some other resistors, so that the reflection coefficient at the right end of line A is ρ_L and the reflection coefficient on the left end is ρ_S. In this case it is best to consider separately the crosstalk due to incident waves and that due to reflected waves.

In the preceding two sections we plotted forward and reverse crosstalk along various positions on the line. In most practical cases it is easiest to observe the waveforms at either end of the victim line, a task that can be difficult to carry out at other locations. As a consequence we will henceforth confine our attention to the crosstalk waveforms at $z = 0$ and $z = l$. An example will best illustrate the method.

Example 5.3 Assume that in figure 5.2 line A is series terminated, hence the right end is terminated in an open circuit while the terminating impedance on the left end of the line is Z_0. Line A is excited by the waveform shown in figure 5.9 and line B is matched at both ends. We wish to determine the crosstalk on line B at $z = 0$ and at $z = l$.

The crosstalk duc to the incident wave was discussed in examples 5.1 and 5.2. In this case we are interested in the solutions at both ends of the line. We simply substitute $z = 0$ and $z = l$ into (5.18) and (5.27) to produce the equations

$$v_c(l, t) = lK_{cf}v_i'(t - T) \tag{5.31}$$

$$v_c(0, t) = K_{cr}[v_i(t) - v_i(t - 2T)] \tag{5.32}$$

Substitution of the incident waveform $v_i(t)$ shown in figure 5.9 into the above equations produces the crosstalk result shown in figure 5.13.

Since the reflection coefficient at the right end of aggressor line A is unity, we know that the reflected wave will be identical to the incident wave except that it will occur T seconds later. Since the reflected wave propagates from right to left the idea of forward and reverse crosstalk on line B is reversed. As a consequence the response that previously occurred at $z = 0$ will now occur at $z = l$ and vice versa. Hence figure 5.14 was obtained from figure 5.13 by simply interchanging the two graphs, and delaying the response by T seconds.

Since the reflection coefficient on the left side of the aggressor line is zero, there are no further reflections and the problem is solved. It only remains to add the solutions at $z = 0$ and $z = l$. ■

We can readily generalize the above analysis to cover additional reflections on line A. We have to simply be aware of the direction of propagation of reflected waves in order to associate the proper response with the appropriate end of the line. It is also important to

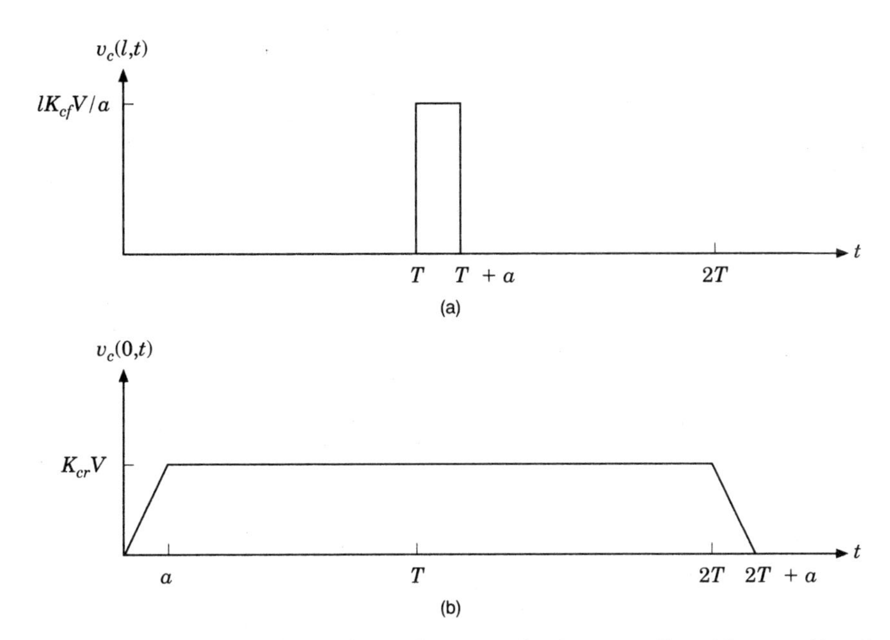

Figure 5.13 The solution due to the incident wave for the crosstalk problem considered in example 5.3.

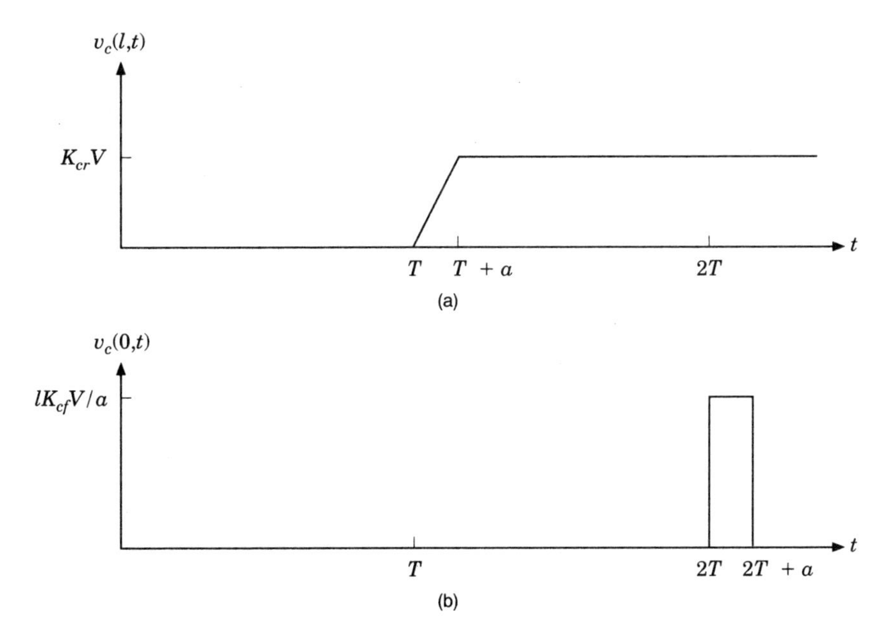

Figure 5.14 The solution due to the reflected wave for the crosstalk problem considered in example 5.3.

account for additional delays in integer multiples of T corresponding to the time that the reflection begins.

5.6 Unmatched Victim Lines

In the last section we considered the case of the unmatched aggressor line. Now we want to consider crosstalk for the case of the unmatched victim line. To facilitate the analysis we will confine ourselves to the case in which one end of the transmission line is terminated in a matched impedance, while the other end has a non-zero reflection coefficient. The method is best presented through an example.

Example 5.4 We will assume that in figure 5.2 the victim line is series terminated, so that its left-hand side has a terminating resistor Z_0, whereas the right side is terminated in an open circuit. Find and plot the crosstalk voltage at $z = 0$ on line B. As in the previous examples, it will be assumed that K_{cf} is positive.

In the discussion that follows it is necessary to refer constantly to figure 5.15. The forward crosstalk waveform increases in size as it progresses from left to right, as was illustrated in example 5.1, and takes on a maximum at $z = l$. The waveform $v_{cf}(l,t)$ can be obtained as the logical continuation of example 5.1 and is illustrated in the top diagram of figure 5.15. The right-hand side of line B is terminated in an open circuit, where the reflection coefficient is unity, hence the incident forward crosstalk voltage $v_{cf}(l,t)$ gives rise to the reflected voltage which is designated $v_{cf-}(l,t)$. This voltage propagates in the direction of decreasing z, and does not change in amplitude, since it is a direct consequence of reflection (and only an indirect consequence of crosstalk). It reaches the left end of line B after a time delay T. This is $v_{cf-}(0,t)$, the third waveform shown in figure 5.15.

From example 5.2 it is clear that the reverse crosstalk waveform $v_{cr}(z,t)$ is zero at the right termination, hence it is not at all affected by the open circuit termination of line B. The waveform, $v_{cr}(0,t)$ shown in figure 5.15, is the reverse crosstalk voltage at $z = 0$ and can be readily obtained by taking example 5.2 to its logical conclusion. The final result shown is $v_{tot}(0,t)$ and was obtained by summing $v_{cf-}(0,t)$ with $v_{cr}(0,t)$. ■

Other problems pertaining to unmatched victim lines can be solved in a manner similar to that used in the above example.

5.7 Measurement of Crosstalk Coefficients

To measure the crosstalk coefficients it is best to try to rely on methods that are very simple in that they do not require very complicated equipment. Sine-wave generators and conventional (non-storage) oscilloscopes are common equipment in most laboratories. The method described below is based on the use of this kind of equipment.

In the setup in figure 5.2, for the source $v_G(t)$ we select a sinusoidal voltage generator whose output is a sine wave of radian frequency

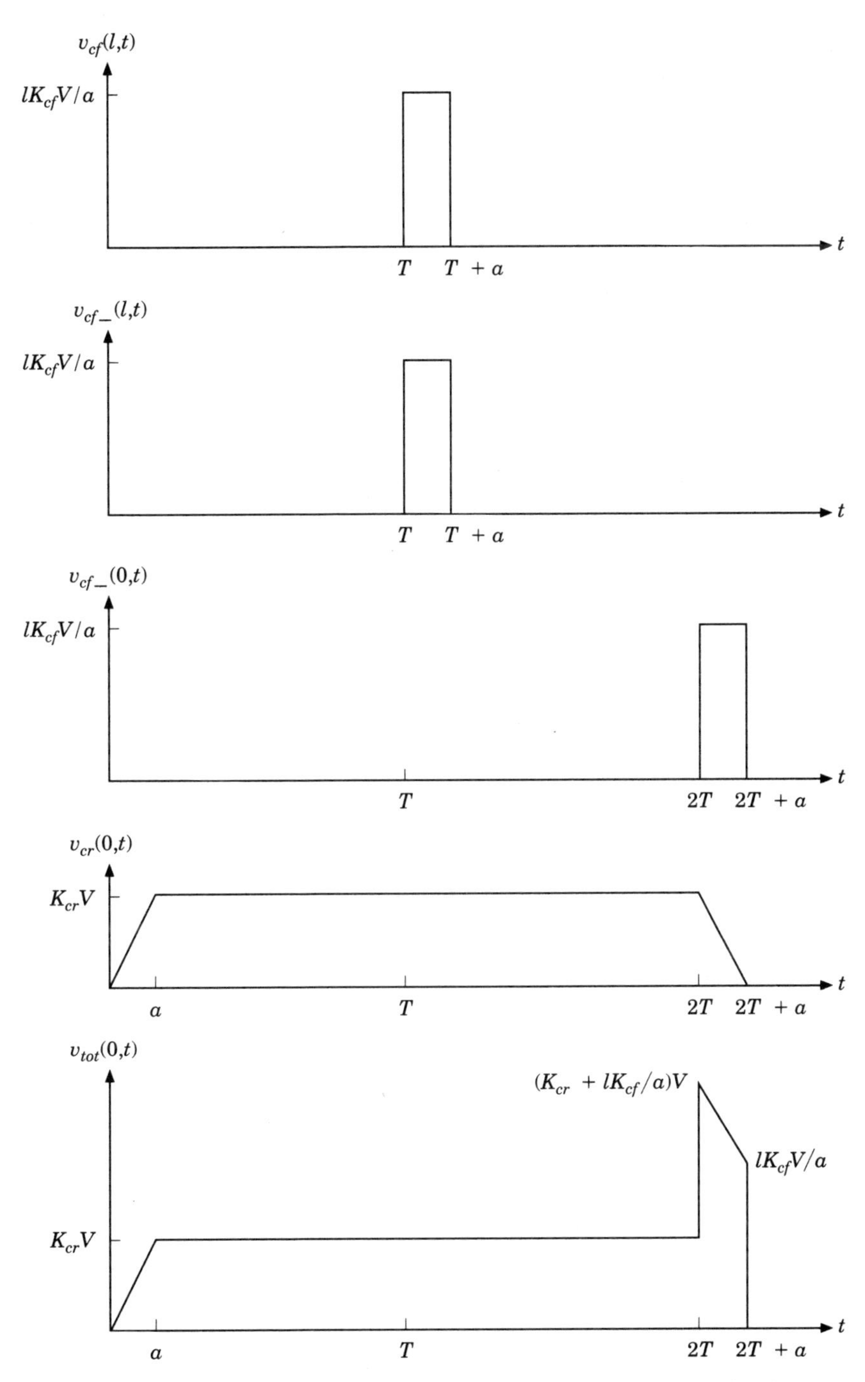

Figure 5.15 The progressive solution to the crosstalk problem considered in example 5.4.

$\omega_0 = 2\pi f_0$. As a consequence, the voltage input to the aggressor line A which is seen on an oscilloscope screen is

$$v_i(t) = V \sin \omega_0 t \tag{5.33}$$

whose peak-to-peak value is given by

$$|v_i(t)|_{pp} = 2V \tag{5.34}$$

Using (5.33) in (5.27), we find that the reverse crosstalk voltage observed on an oscilloscope at the left end of victim line B will be

$$v_{cr}(0, t) = K_{cr}\{V \sin \omega_0 t - V \sin [\omega_0(t - 2T)]\} \tag{5.35}$$

We wish to avoid having to determine accurately the product $\omega_0(2T)$ needed in the above expression, and at the same time to improve the observability of the reverse crosstalk waveform. Accordingly we adjust the frequency of the sine-wave generator $v_G(t)$ so that waveform $v_{cr}(0, t)$, which is displayed on the oscilloscope, is maximized. This occurs when

$$\omega_0(2T) = \pi \tag{5.36}$$

or

$$f_0 = \frac{1}{4T} \tag{5.37}$$

When the above condition obtains, the observed reverse crosstalk is given by

$$v_{cr}(0, t) = 2K_{cr} V \sin \omega_0 t \tag{5.38}$$

whose peak-to-peak voltage is

$$|v_{cr}(0, t)|_{pp} = 4K_{cr} V \tag{5.39}$$

The forward crosstalk voltage is observed on an oscilloscope at the right end of the victim line. Substituting (5.33) into (5.18) we find the expression for this voltage to be

$$v_{cf}(l, t) = l\, \omega_0 K_{cf} V \cos[\omega_0(t - T)] \tag{5.40}$$

Using the condition from (5.36) we reduce the above to

$$v_{cf}(l, t) = l\, \omega_0 K_{cf} V \sin \omega_0 t \tag{5.41}$$

which has a peak-to-peak value given by

$$|v_{cf}(l,t)|_{pp} = 2l\,\omega_0|K_{cf}|V \tag{5.42}$$

The absolute magnitude of K_{cf} is used above because the forward crosstalk coefficient can be either positive or negative, as is readily apparent from (5.9).

Taking suitable ratios, we find that the crosstalk coefficients can be determined using

$$K_{cr} = \frac{|v_{cr}(0,t)|_{pp}}{2|v_i(t)|_{pp}} \tag{5.43}$$

and

$$|K_{cf}| = \frac{|v_{cf}(l,t)|_{pp}}{l\,\omega_0|v_i(t)|_{pp}} \tag{5.44}$$

To determine the sign of K_{cf} we observe that $v_{cf}(l,t)$ in (5.41) is in phase with $v_i(t)$ in (5.33) if K_{cf} is positive. If the two voltages are 180° out of phase, then this indicates that K_{cf} is negative.

Example 5.5 Two closely-spaced transmission-lines have a common length $l = 10\,\text{m}$ and a one-way delay $T = 50\,\text{ns}$. From (5.37) we find that the proper frequency for performing the test is $f_0 = 5$ MHz. Accordingly the voltage generator is adjusted, as accurately as possible, to produce a sine wave of 5 MHz. The reverse crosstalk voltage observed at the left end of the victim line B is maximized by fine tuning the frequency adjusting dial on the sine-wave generator.

It is found, by observing $v_i(t)$ on a well-calibrated oscilloscope, that indeed $f_0 = 5$ MHz. It is also found that $|v_i(t)|_{pp} = 10\,\text{V}$. In addition it is observed that $|v_{cr}(0,t)|_{pp} = 0.3\,\text{V}$ and $|v_{cf}(l,t)|_{pp} = 0.0315\,\text{V}$. Furthermore, it is found that $v_{cf}(l,t)$ is 180° out of phase with $v_i(t)$ which indicates immediately that K_{cf} is negative.

From the data given we readily find by using (5.43) and (5.44) that $K_{cr} = 0.015$ and $K_{cf} = -0.01\,\text{ns/m}$. ■

5.8 Conclusion

Crosstalk is of great concern in the design of interconnections in digital computer applications. One might be tempted to try to keep signal rise time small, in order to minimize forward crosstalk, but this choice does not always exist. Usually the drive signal is determined by the choice of the type of digital logic-devices. This specifies the drive signal, which determines the forms of $v_i(\cdot)$ and $v_i'(\cdot)$. To keep crosstalk coupling to a minimum, it is necessary to observe some precautions.

- The length of printed circuit board traces over which adjacent signal carrying lines are parallel should be kept to a minimum.
- If lines must run parallel, then they should be well separated so that the side-to-side capacitance of the lines will be small.
- Adjacent signal layers in printed circuit boards should be separated by ground planes.
- Adjacent signal-carrying conductors should be separated with conductors grounded at both ends. This is particularly feasible in the case of flat parallel cables. An example of this is the cable used to connect the parallel ports of personal computers to printers.
- Use of twisted-pair transmission-lines confines the electromagnetic field largely to the two wires and has a tendency to cancel signals induced from other lines, because both wires pick up essentially the same signal.
- Where possible use coaxial cable. Coaxial cable confines almost the entire electromagnetic field to the region between the two conductors. The field radiated to the space outside the cable is minimal and conversely the coaxial cable picks up very little signal from other circuits.

Bibliography

1. B. Hart, *Digital Signal Transmission — Line Circuit Technology* (Chapter 6), Van Nostrand Reinhold Co. Ltd., Wokingham, U.K., 1988.
2. A. Feller, H. R. Kaupp, and J. J. Digiacomo, "Crosstalk and Reflections in High-Speed Digital Systems, *Proceedings — Fall Joint Computer Conference*, 1965, pp. 511–525.
3. *High Speed Digital Symposium*, held at Hewlett-Packard, Paramus, New Jersey, August 8, 1990. Valuable notes were made available. Of particular interest were those by Dr. Edward P. Sayre of North East Systems Associates, 256 Great Road—Suite 13, Littleton, Massachusetts.

Problems

P5.1 In the paragraph above (5.2) the statement is made that "Strictly speaking, the negative end of the current source should be shown attached to the upper conductor of transmission line A to be in agreement with figure 5.3. But the solution to the problem is not at all affected by showing the negative end of the current source attached to the ground conductor of transmission line B." Convince yourself by examples that this is indeed the case.

P5.2 How can (5.9) and (5.10) be modified to account for the fact that the two interacting transmission-lines possess different characteristic-impedances?

Hint: The solution to this problem requires a re-examination of the deriving (preceding) equations.

P5.3 For example 5.1 find the expression and plot the forward crosstalk at the receiving (right) end of the line.

P5.4 For example 5.2 find the expression and plot the reverse crosstalk at the sending (left) end of the line.

P5.5 Determine the following:

(a) Where is the point on the line where we have reverse crosstalk exclusively?

(b) Where is the point on the line where we have forward crosstalk exclusively?

(c) Consider the answers to parts (a) and (b). Can a setup as shown in figure 5.2 be used to measure the forward and reverse crosstalk coefficients by using the waveform of figure 5.9 for excitation?

P5.6 A raised cosine pulse, described by

$$v_i(t) = V(1 - \cos 2\pi f_0 t), \qquad \text{for } 0 \leq t \leq 1/f_0$$

is introduced on the left end of the aggressor line shown in figure 5.2. Assume that K_{cf} is positive and $T \geq 1/(2f_0)$. Plot $v_{cf}(l,t)$ and $v_{cr}(0,t)$ which appear at the ends of the victim line.

P5.7 Suppose in figure 5.2 the victim line has length l but the aggressor line is longer than l. In what way, if any, would that affect the results?

P5.8 Suppose in figure 5.2 the aggressor line has length l but the victim line is longer than l. In what way, if any, would that affect the results?

P5.9 In example 5.3, find the total voltages $v_{tot}(0,t)$ and $v_{tot}(l,t)$ from the waveforms plotted in figures 5.13 and 5.14.

P5.10 Suppose in figure 5.2 the aggressor line is terminated in a short circuit on the right and is matched with a resistance Z_0 on the left end. Reanalyze example 5.3.

P5.11 Suppose in example 5.4 the victim line is terminated in a short circuit on the right and in Z_0 on the left. Reanalyze example 5.4 to find and plot $v_{tot}(0,t)$.

P5.12 Repeat example 5.4 assuming that line B is parallel terminated, namely the right-hand side is terminated in Z_0, whereas the left-hand side is terminated in a short circuit. Solve for $v_{tot}(l,t)$.

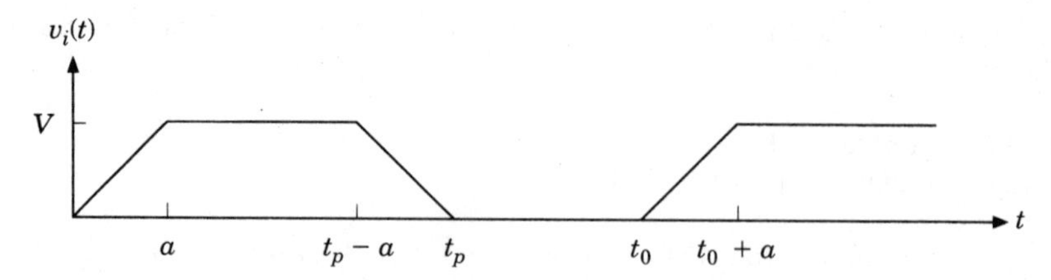

Figure 5.16 The input voltage on aggressor line A for measuring crosstalk coefficients.

P5.13 In connection with (5.36), the desired conditions could be met with $\omega_0(2T)$ equal to any odd multiple of π. Why are the other choices for $\omega_0(2T)$ less desirable for this test?

P5.14 In attempting to measure the crosstalk coefficients using the method described in section 5.7, it is found that the sine-wave generator has a DC offset. In place of (5.33) we actually have

$$v_i(t) = V \sin \omega_0 t + V_{DC}$$

Will this affect the experiment? Discuss in detail.

P5.15 We have the same specifications as in example 5.5, but the lines are more closely spaced so that the crosstalk voltages are larger. The voltages observed are $|v_i(t)|_{pp} = 10\,\text{V}$, $|v_{cr}(0,t)|_{pp} = 0.7\,\text{V}$, and $|v_{cf}(l,t)|_{pp} = 0.043\,\text{V}$. It is also noted that $v_{cf}(l,t)$ is in phase with $v_i(t)$. Find the crosstalk coefficients for this case.

P5.16 A voltage wave-generator is available that can produce the periodic waveform shown in figure 5.16. This is to be used as the line input voltage $v_i(t)$.

(a) Sketch the crosstalk waveforms at $z = 0$ and $z = l$ that will be observed on an oscilloscope.

(b) To keep the oscilloscope observations simple, it is best to adjust the waveform parameters so that the crosstalked pulses will not overlap. Carefully determine the relation between t_p, t_0 and the transmission-line transit-time T, so that the above requirement will be satisfied.

(c) Write the equations for determining K_{cr} and K_{cf} that replace (5.43) and (5.44). What is the condition that determines if K_{cf} is positive or negative?

Chapter

6

Interconnecting High Speed ECL

6.1 Introduction

In this chapter we will apply the principles of the earlier chapters to solve some interconnection problems that arise when dealing with fast logic devices. The methods will be demonstrated on the 100K series of emitter-coupled logic (ECL) devices which are among the fastest logic circuitry available at this time. This in no way means that this material cannot be generalized, by the simple changing of device parameters, to apply to the faster gallium-arsenide (GaAs) devices. These are now becoming popular and threaten to replace ECL as the fastest digital technology available.

The propagation delay of a signal through some types of 100K gates is less than 1 ns. A signal traveling 33 cm ($\approx$ 1 foot) in free air is delayed 1 ns. On the printed circuit (PC) boards used for connecting the logic devices, signals travel slower than they would through air, so that even for a moderately sized board, for example one that is 6 inches (or 15 cm) square, the effects of transmission line phenomena on the digital signals must be considered. It is therefore necessary to take PC board delays into consideration when dealing with fast logic circuits. At this point a review of some of the ECL material presented in appendix A is appropriate.

6.2 Modeling of ECL Circuits

The model to use for analyzing the problem of determining how to terminate the transmission lines interconnecting the ECL circuitry is based on the material found in appendix A. The input circuitry of ECL is modeled as a 50 kΩ resistance in parallel with a parasitic 2 pF capacitance, so it should be kept in mind that every transmission line that connects logic devices will have a capacitive termination. The input voltages that are of interest in the case of 100K ECL are the ones that appear in table A.1. They are reproduced here for the reader's convenience in table 6.1.

In connection with table 6.1, we have the following two definitions. The amount by which the high state input signal V_{IH} exceeds the minimum required voltage V_{IHmin} is the high state noise margin V_{NH}, given by

$$V_{NH} = V_{IH} - V_{IHmin} \tag{6.1}$$

The amount by which the low state input signal V_{IL} falls below the maximum required voltage V_{ILmax} is the low state noise margin V_{NL}, given by

$$V_{NL} = V_{ILmax} - V_{IL} \tag{6.2}$$

The model for the output circuitry of 100K ECL will be based on the diagram of figure A.3 of appendix A.

6.3 Model for Line with Initial Conditions

Consider the transmission line in figure 6.1. When the switch is in position 1, the transmission line attains a specific voltage and a specific current just prior to the time the switch is thrown. These two quantities constitute the initial conditions for the transmission line just after the switch is thrown into position 2. We will refer to these initial conditions as the voltage $V(0+)$ and the current $I(0+)$. After the switch makes the transition, the voltage $V(0+)$ and the current $I(0+)$ no longer satisfy the circuit conditions at the sending end. As indicated

TABLE 6.1 Some critical values for the input voltage V_I for 100K ECL.

V_I	Gate input-voltage
$V_{IHmax} = -0.880\,\text{V}$	Saturation of an input transistor risked above this value of V_I.
$V_{IHmin} = -1.165\,\text{V}$	All manufactured units will make a transition to the high state when the input voltage exceeds this value.
$V_{ILmax} = -1.475\,\text{V}$	All manufactured units will make a transition to the low state when the input voltage falls below this value.
$V_{ILmin} = -1.810\,\text{V}$	Saturation of an input transistor risked below this value of V_I.

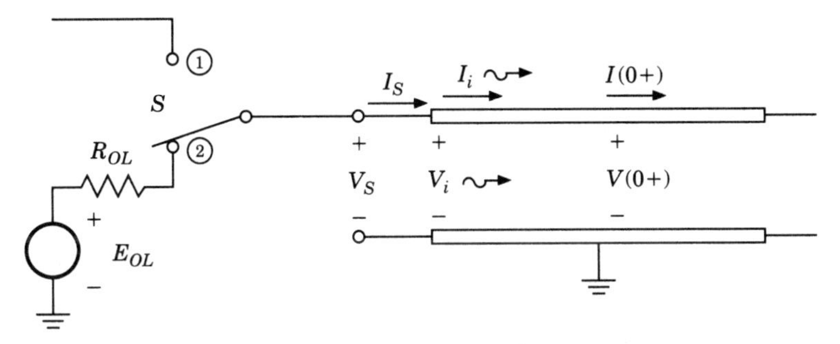

Figure 6.1 A transmission line with a switched source.

in figure 6.1, the line responds by sending an incident voltage wave V_i and an incident current wave I_i, which we have found in previous chapters are related by

$$V_i = Z_0 I_i \tag{6.3}$$

We wish to determine the incident voltage V_i and the sending end voltage V_S immediately after the switch is thrown to position 2. The new sending end voltage V_S, immediately after switching, is

$$V_S = V(0+) + V_i \tag{6.4}$$

and the sending end current is

$$I_S = I(0+) + I_i \tag{6.5}$$

The circuit requirement that must be satisfied is

$$V_S + I_S R_{OL} = E_{OL} \tag{6.6}$$

Substituting (6.3), (6.4), and (6.5) into (6.6) and solving for I_i we get

$$I_i = \frac{E_{OL} - V(0+)}{R_{OL} + Z_0} - \frac{R_{OL}}{R_{OL} + Z_0} I(0+) \tag{6.7}$$

We note in passing that once we know the incident current I_i, the incident voltage V_i can be readily found using (6.3), and the sending end voltage V_S can then be found from (6.4).

Eliminating I_i from (6.7) and (6.5) and solving for the sending end current I_S we get

$$I_S = \frac{E_{OL} - V(0+)}{R_{OL} + Z_0} + \frac{Z_0}{R_{OL} + Z_0} I(0+) \tag{6.8}$$

Examination of the above expression for sending-end current reveals that it could have been derived from the lumped parameter model depicted in figure 6.2. The two sources shown in the model are due to the initial conditions imposed on the transmission line before the switch is thrown from position 1 to position 2. This model will be very useful in the subsequent material.

6.4 Termination Requirements of Logic Circuit PC Boards

It has been mentioned in previous chapters that the printed circuit (PC) board traces used to interconnect digital logic devices constitute transmission lines, hence proper attention has to be given to their termination. When signals propagate on a transmission line, substantial reflections from unterminated lines are possible. It is best either to terminate the transmission line in its characteristic impedance, or to terminate it in such a way that the signal, with its reflections, will still produce desirable voltage levels at the end of the transmission line where the inputs of logic gates are found.

For ECL, there are a number of methods of terminating PC board circuits to obtain proper functioning. The first two, parallel termination and series termination, have already been mentioned in

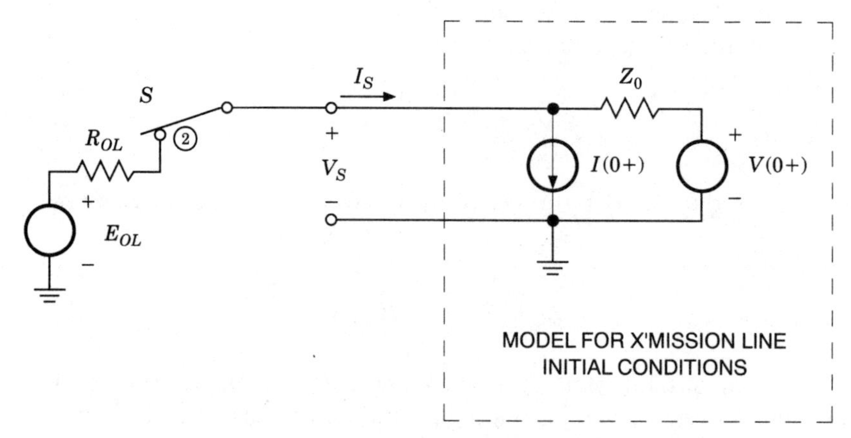

Figure 6.2 A lumped-parameter model with transmission-line initial-conditions.

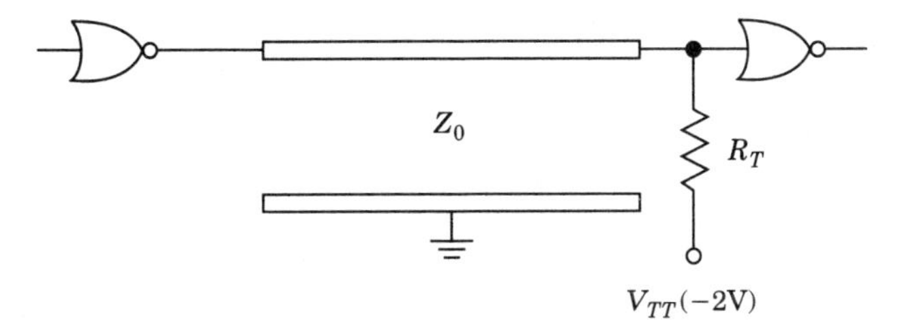

Figure 6.3 Parallel-terminated ECL.

chapter 2. It is also possible to leave the lines unterminated when the transmission-line connections are very short. The various methods of termination will be discussed in the following sections.

6.5 Parallel Termination

Terminating the transmission line at the receiving end with a resistor R_T, as shown in figure 6.3, is referred to as parallel termination. The value of R_T is chosen so that it will function as both a pulldown resistor (whose need for ECL logic devices is discussed in appendix A), and to suppress reflections from the receiving end of the transmission line. It is easiest to simply choose R_T to equal Z_0. This way any voltage wave launched at the transmitting end would be totally absorbed at the receiving end, since the line is matched there, and the transmission process would terminate. But it is often desired to reduce the power dissipation in the resistor R_T, so a value higher than Z_0 is often chosen, and some reflections may have to be accepted. Before proceeding with the discussion, we replace the output gate shown on the left of figure 6.3 with the idealized equivalent circuit of figure A.3 of appendix A, with the result shown in figure 6.4.

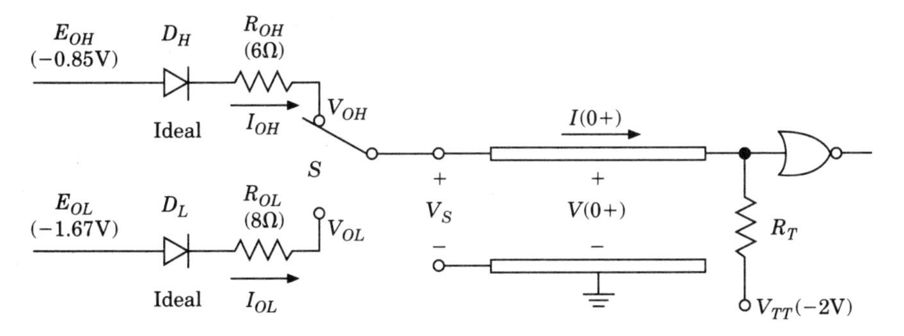

Figure 6.4 Figure 6.3 with an idealized model of the gate output. The numbers shown are for 100K ECL.

We wish to find a value of R_T that will be higher than Z_0, to minimize power dissipation, but not so high that the functioning of the receiving gates will be impaired due to an improper voltage at the gate input. One possible criterion for selection is to choose R_T to be the highest value that will still keep the emitter of the output gate conducting at all times. This is equivalent to keeping the ideal diode D_L in figure 6.4 conducting after the switch is thrown to the lower position. The reasons for choosing this criterion will become apparent later.

To derive the requirements on the resistor R_T, it will be assumed the transmission system shown in figure 6.4 has reached equilibrium in the high output state. We readily determine that the steady-state current along the entire line is

$$I_{OH} = I(0+) = \frac{E_{OH} - V_{TT}}{R_{OH} + R_T} \tag{6.9}$$

and the steady-state voltage along the entire transmission-line is given by

$$V_{OH} = V(0+) = \frac{R_T}{R_{OH} + R_T} E_{OH} + \frac{R_{OH}}{R_{OH} + R_T} V_{TT} \tag{6.10}$$

After the gate output goes to the low state, we are dealing with the circuit of figure 6.4 with the switch in the lower position. To facilitate the analysis we use figure 6.2 to redraw figure 6.4 in the lumped parameter equivalent model of figure 6.5. The resulting diagram is used to solve easily for V_{OL} for two cases.

If V_{OL} is below E_{OL} then diode D_L is conducting, and we have

$$V_S = V_{OL} = \frac{Z_0}{R_{OL} + Z_0} E_{OL} + \frac{R_{OL}}{R_{OL} + Z_0} V(0+) - \frac{R_{OL} Z_0}{R_{OL} + Z_0} I(0+), \quad \text{for } V_{OL} < E_{OL} \tag{6.11}$$

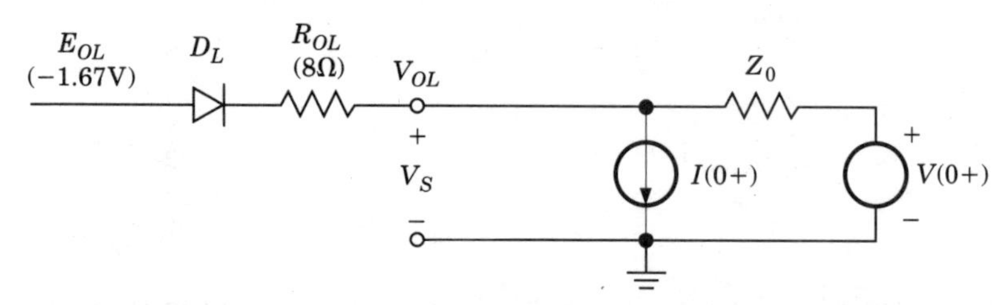

Figure 6.5 Figure 6.4 with lumped-parameter transmission-line model.

If V_{OL} is greater than or equal to E_{OL}, then diode D_L is cut off and we get the much simpler form

$$V_S = V_{OL} = V(0+) - Z_0 I(0+), \qquad \text{for } V_{OL} \geq E_{OL} \tag{6.12}$$

It is best to keep the ECL output-transistor emitter conducting at all times, so that the transmission line will always be driven by a low impedance source. The above condition is equivalent to requiring that ideal diode D_L conduct at all times. The desired range of values for R_T that achieves this condition can be found by demanding that either the value of V_{OL} given in (6.11) be less than E_{OL}, or that R_T take on values smaller than the value needed to have V_{OL} in (6.12) just equal E_{OL}. Using (6.9) and (6.10) to eliminate $I(0+)$ and $V(0+)$ in the above two equations, and then proceeding to solve for R_T, we get by either method

$$R_T < \frac{(E_{OH} - V_{TT})Z_0 + (E_{OL} - V_{TT})R_{OH}}{E_{OH} - E_{OL}} \tag{6.13}$$

The values shown in figure 6.4 are for 100K ECL, and for that case (6.13) evaluates to

$$R_T < 1.4\,Z_0 + 2.41 \tag{6.14}$$

It will now be demonstrated through examples that it is best to choose the parallel termination-resistor R_T in accordance with (6.13).

Example 6.1 In this example we will deal with a 100K ECL gate working into a 50 Ω transmission line. The model used is that of figure 6.4. $R_T = 50\,\Omega$ is used for a matched parallel-termination. Using (6.9) and (6.10) we obtain

$$I_{OH} = I(0+) = 20.54\,\text{mA}$$

$$V_{OH} = V(0+) = -0.973\,\text{V}$$

From (6.11) we also find that D_L is conducting since

$$V_S = V_{OL} = -1.716\,\text{V} < -1.67\text{V} = E_{OL}$$

In order to draw the lattice diagram we need V_i, which we find by substituting the above value of V_S into (6.4) to obtain

$$V_i = -0.742\,\text{V}$$

The line is matched at the receiving end so there will be no reflections. The lattice diagram and the voltage at the load are shown in figure 6.6 for a summary

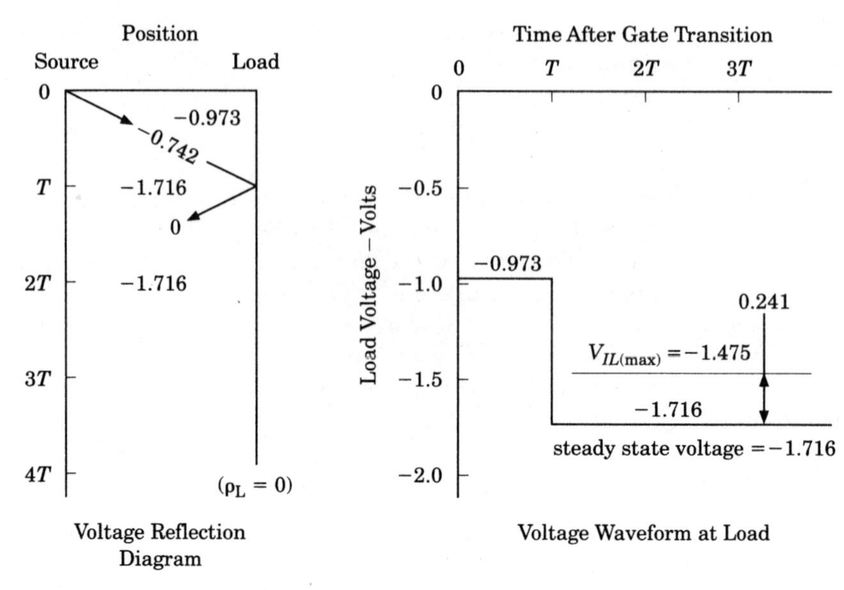

Figure 6.6 Summary of the circuit behavior for 100K ECL operating with a 50 Ω transmission line with a parallel termination of 50 Ω.

of the circuit behavior. From these diagrams we see that the voltage on the line reaches its final value upon the arrival of the first incident wave. A gate connected to the end of the line must wait T seconds for the voltage to reach its final value, whereas gates distributed along the line need to wait less. This is the distinct advantage of using a matched parallel-termination. Because this is the lowest possible value of R_T, it has the highest power dissipation. Since the voltage at the receiving gate undershoots V_{ILmax} by 241 mV, we can say the noise margin at the receiving gate during negative transitions is 241 mV. ■

Example 6.2 The same data as in example 6.1 but in this case the resistance of R_T used for a parallel termination is 70 Ω. This value is just a trifle below the upper bound of (6.14). As before, we use (6.9) and (6.10) to obtain

$$I_{OH} = I(0+) = 15.13\,\text{mA}$$

$$V_{OH} = V(0+) = -0.941\,\text{V}$$

We find from (6.11) that

$$V_S = V_{OL} = -1.674\,\text{V} < -1.67\text{V} = E_{OL}$$

and we have verified that diode D_L is indeed forward biased. For the purpose of drawing the lattice diagram we need V_i, which we find by substituting the above value of V_S into (6.4) to obtain

$$V_i = -0.733\,\text{V}$$

The line is not matched at the receiving end so there will be some reflections. The reflection coefficient at the receiving end is $\frac{1}{6}$. The lattice diagram and the voltage at the load are shown in figure 6.7 for a summary of the circuit behavior. The voltage at the receiving gate undershoots the final value of $-1.704\,\mathrm{V}$ by approximately 90 mV. The remaining oscillations are relatively insignificant in value. The highest voltage value at the receiving gate is $-1.693\,\mathrm{V}$ and occurs after $3T$ seconds, so the smallest noise margin for negative transitions is 218 mV. Any gates that are distributed along the line will experience an earlier initial voltage change to $-1.674\,\mathrm{V}$ as the first incident wave passes them, which is different from the situation encountered at the load, where the voltage changes to $-1.796\,\mathrm{V}$ after T seconds. ■

Example 6.3 The same data as in example 6.1 but in this case the value of R_T used for a parallel termination is $100\,\Omega$. This value is well above the upper bound of (6.14), so we can expect diode D_L to be reverse biased immediately after the output-gate voltage-transition. As in the previous examples, we use (6.9) and (6.10) to obtain

$$I_{OH} = I(0+) = 10.85\,\mathrm{mA}$$

$$V_{OH} = V(0+) = -0.915\,\mathrm{V}$$

Since D_L is very likely to be initially reverse biased, we use (6.12) to find

$$V_S = V_{OL} = -1.458\,\mathrm{V} > -1.67\mathrm{V} = E_{OL}$$

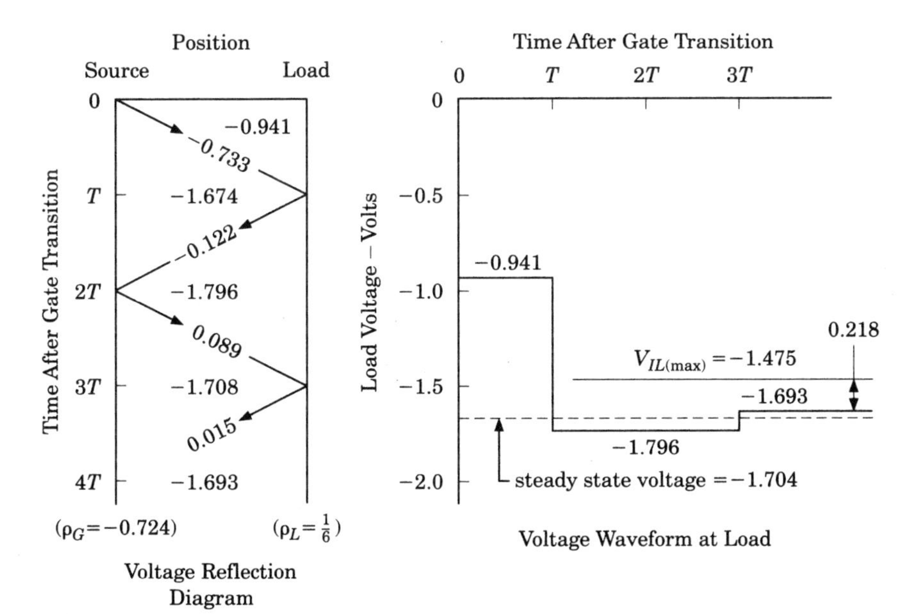

Figure 6.7 Summary of the circuit behavior for 100K ECL operating with a $50\,\Omega$ transmission line with a parallel termination of $70\,\Omega$.

and we have verified that diode D_L is indeed initially turned off. For purposes of drawing the lattice diagram we need V_i, which we find by substituting the above value of V_S into (6.4) to obtain

$$V_i = -0.542\,\text{V}$$

The line is distinctly mismatched at the receiving end so there will be some reflections. The reflection coefficient at the receiving end is 1/3. The lattice diagram and the voltage at the load are shown in figure 6.8 for a summary of the circuit behavior.

To arrive at the reflection diagram, we take into account the fact that at the sending end the voltage V_S is above the voltage E_{OL} until time $2T$. Therefore the ideal diode D_L is initially turned off and the reflection coefficient at the sending end is +1 up to time $2T$, when the first reflected wave arrives from the receiving end. But as soon as the first reflection arrives at the sending end of the line at time $2T$, the voltage V_S drops to $-1.819\,\text{V}$, causing the diode D_L to turn on after a short, but indeterminate, time delay of duration Δ. With D_L in conduction, the voltage-current conditions at the sending end of the line are no longer satisfied, so a second incident wave is launched at time $2T + \Delta$.

It is readily apparent from figure 6.4 that with the output gate in the low-logic state, any current that exists on the transmission line must pass through the ideal diode D_L. Since D_L is cut off up to time $2T + \Delta$, then the current $I(0+)$, which we will use as an initial condition, is zero. The voltage representing the initial condition prior to diode turn-on is found from the lattice diagram

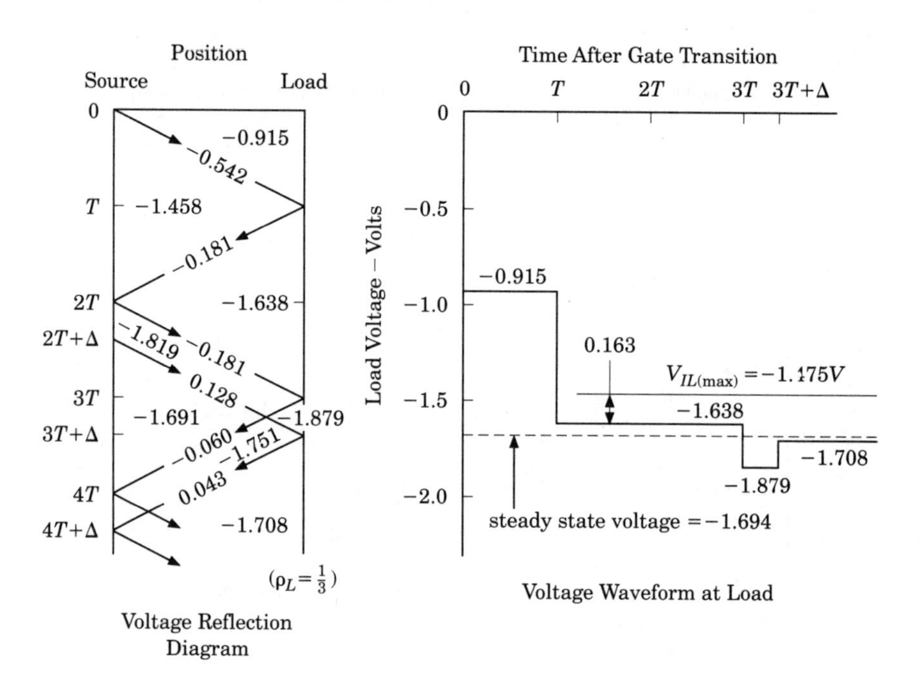

Figure 6.8 Summary of the circuit behavior for 100K ECL operating with a 50 Ω transmission line with a parallel termination of 100 Ω.

an instant after $2T$ seconds. The value of $V(0+)$ is $-1.819\,\mathrm{V}$, hence (6.7) evaluates to

$$I_i = \frac{E_{OL} - V(0+)}{R_{OL} + Z_0} = \frac{-1.67 - (-1.819)}{8 + 50} = 2.57\,\mathrm{mA}$$

Since the incident voltage and current are related by Z_0, we get V_i by multiplying the above by Z_0 to obtain

$$V_i = 0.128\,\mathrm{V}$$

The progress of the second incident voltage has to be traced in the lattice diagram in order to obtain a complete analysis. It can be readily seen from the lattice diagram that after the time $2T + \Delta$ the diode D_L remains turned on, so no further complications in the analysis take place. But there are other problems with the circuit when the value of R_T violates the inequality of (6.13). They are listed below.

- At $3T$ seconds, the voltage at the receiving gate initially undershoots the final value of $-1.694\,\mathrm{V}$ by approximately 185 mV. The remaining oscillations are very small to be of significance. The highest voltage value at the receiving gate is $-1.638\,\mathrm{V}$ and occurs after T seconds, so the smallest noise margin at the load for negative transitions is 163 mV which is less than that calculated in example 6.2.
- Any gate inputs that are distributed along the length of the line will experience an initial voltage change to $-1.458\,\mathrm{V}$ which exceeds the voltage $V_{ILmax} = -1.475\,\mathrm{V}$. So there is no assurance that those gates will change state.
- It is readily apparent from the lattice diagram that any gate inputs that are distributed along the length of the line will experience at their input a voltage of $-1.819\,\mathrm{V}$ for Δ seconds, and the voltage at the load will be $-1.879\,\mathrm{V}$ for Δ seconds. The input circuits of all the above-mentioned gates will go into saturation. This is very undesirable for ECL since it causes a slowing down in their operation.

It can be said in summary that it is best to choose a value of R_T that satisfies (6.13) unless there are very good reasons for not doing so. ■

6.6 Parallel-Terminated Multiple-Gates

We have seen in examples 6.1 and 6.2 that for a high-to-low signal transition, with proper parallel-termination, any gate inputs connected along the middle portion of the transmission line get a sufficiently low input voltage during the propagation of the first incident wave. It is therefore possible to connect gates in the manner shown in figure 6.9. The limiting factors will be discussed very briefly.

An output gate can deliver a maximum current of 50 mA in the high state [1,2]. In the low state each gate draws an input current which is given by

$$I_{IL} = 0.5\,\mu\mathrm{A} \tag{6.15}$$

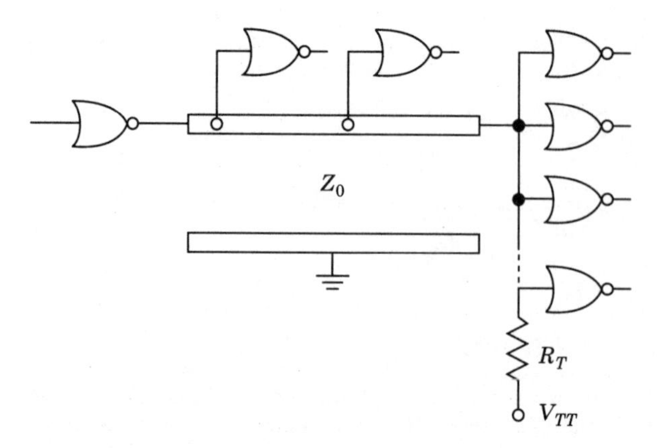

Figure 6.9 A possible connection of gates to a parallel-terminated line.

In the high state each gate input draws a current which can be as high as I_{IH}, given by

$$I_{IH} = 350\,\mu\text{A} \tag{6.16}$$

In the low state the current I_{IL} taken by each gate lowers V_{OL} and therefore increases the noise margin. Since the current is drawn through an $8\,\Omega$ resistance, the margin improves by 4 mV for each gate that is driven. This improvement is insignificant and is in severe contrast to what happens in the high state.

In the high state the current I_{IH} is drawn through a $6\,\Omega$ resistance, so the noise margin is reduced by as much as 2.1 mV for each driven gate. This is still very small when compared to the situation that exists for series termination. There, the noise reduction is an order of magnitude higher, as we will see in the next section.

For parallel terminations in the high state, each gate input may take as much as $350\,\mu$A of current. This is not likely to be a limitation since many gates are needed to exceed the 50 mA gate-output current-rating. The resistor R_T connected to a -2 V supply will take approximately 20 mA if its value is only $50\,\Omega$, so this may account for the largest portion of the 50 mA gate output current.

The increase in the signal rise-time due to the combined input capacitance of many loading gates connected in parallel is a more serious consideration, and its effects should be carefully assessed by using the analysis of chapter 3.

6.7 Series Termination

Terminating the transmission line at the sending end as shown in figure 6.10 is referred to as series termination. In this type of

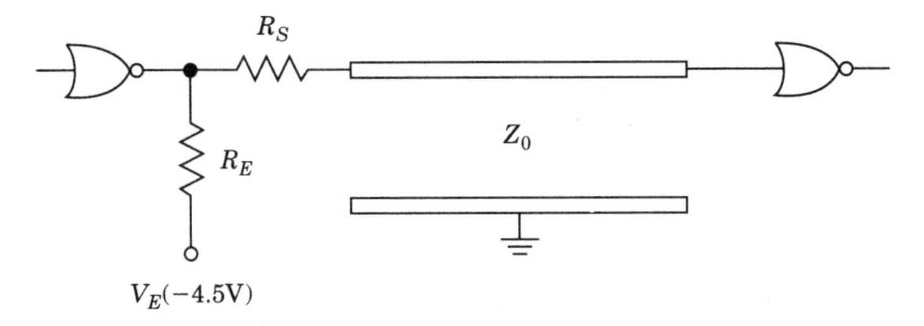

Figure 6.10 Series terminated ECL.

termination the pulldown resistor R_E is placed at the output of the gate that is originating the digital signals. The pulldown voltage V_{EE} is the same as is used to properly bias 100K ECL, usually $-4.5\,V$. In addition, a resistor R_S is placed in series with the sending end of the transmission line to effect an impedance match.

As a first step we redraw the circuit of figure 6.10 by using the idealized model of the gate output circuit of figure A.3 to obtain figure 6.11. The resistor R_S is in series with the transmission line to absorb reflected signals returning toward the sending end. It will be found later that the pulldown resistor R_E is much bigger than either of the gate output resistance values R_{OH} or R_{OL}. In view of this fact, R_E will be ignored when the value of R_S is selected. To perform its function well in both the high and the low state, a good compromise choice for R_S is

$$R_S = Z_0 - \frac{R_{OL} + R_{OH}}{2} \tag{6.17}$$

To derive the requirements on the resistor R_E, we will assume the transmission system shown in figure 6.11 has reached equilibrium in the high output state. If we assume that the input current drawn by

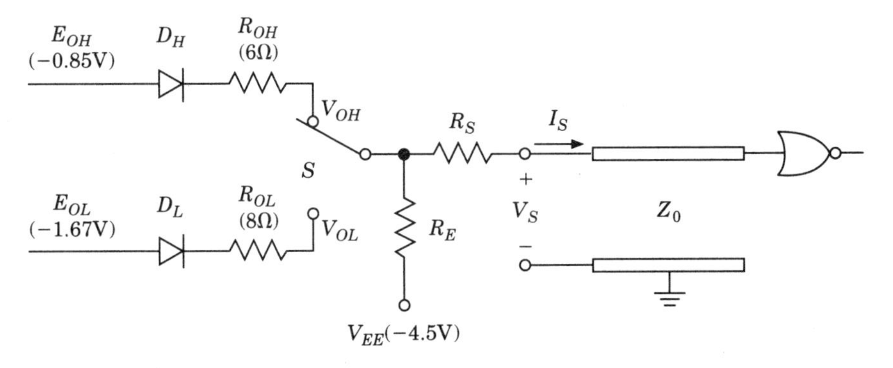

Figure 6.11 Figure 6.10 with idealized output model.

the load gate is inconsequential, then it is clear that the current on the transmission line before switching takes place is zero. The steady-state voltage along the entire transmission-line is given by

$$V_{OH} = V(0+) = \frac{R_E}{R_{OH} + R_E}E_{OH} + \frac{R_{OH}}{R_{OH} + R_E}V_{EE} \qquad (6.18)$$

After the gate output goes to the low state, we are dealing with the circuit of figure 6.11 with the switch in the lower position. To facilitate the analysis, we replace the transmission line in figure 6.11 with the lumped parameter model containing initial conditions, as shown in figure 6.12.

We can solve for V_{OL} (using the superposition principle) for two cases. If V_{OL} is below E_{OL}, then diode D_L is conducting, and we have after some simplification

$$V_{OL} = \frac{R_E(R_S + Z_0)E_{OL} + R_{OL}R_EV_{OH} + R_{OL}(R_S + Z_0)V_{EE}}{(R_{OL} + R_E)(R_S + Z_0) + R_{OL}R_E}, \quad \text{for } V_{OL} < E_{OL} \qquad (6.19)$$

If V_{OL} is greater than or equal to E_{OL}, then diode D_L is cut off and we get

$$V_{OL} = \frac{R_E}{R_E + R_S + Z_0}V_{OH} + \frac{R_S + Z_0}{R_E + R_S + Z_0}V_{EE}, \qquad \text{for } V_{OL} \geq E_{OL} \qquad (6.20)$$

The desired range of values for R_E that we seek can be found by either demanding that the value of V_{OL} given in (6.19) be less than E_{OL}, or that R_E take on values smaller than the value R_{Emax} needed to have V_{OL} in (6.20) just equal E_{OL}. To avoid a great deal of tedious

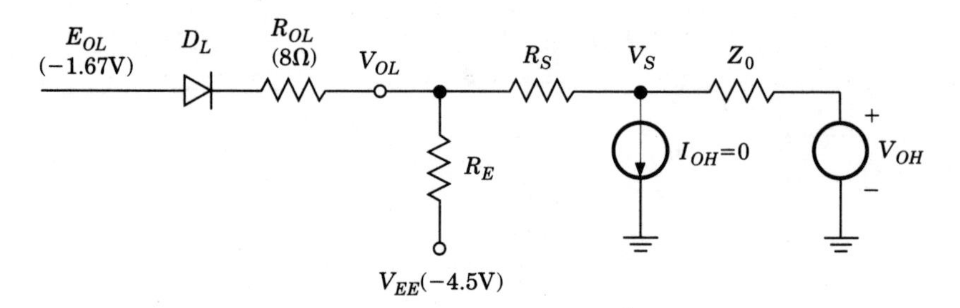

Figure 6.12 Figure 6.11 with lumped-parameter transmission-line model.

algebra by working with (6.19), we eliminate V_{OH} in (6.20) by using (6.18), then set V_{OL} equal to E_{OL} and proceed to solve for R_E. This results in the quadratic equation for R_{Emax}

$$(E_{OH} - E_{OL})R_{Emax}^2 - (Z_0 + R_S + R_{OH})(E_{OL} - V_{EE})R_{Emax} - (Z_0 + R_S)(E_{OL} - V_{EE})R_{OH} = 0 \quad (6.21)$$

For the numbers shown in figure 6.11 for 100K ECL, with R_S chosen using (6.17), (6.21) reduces to

$$R_{Emax}^2 - 3.45(2Z_0 - 1)R_{Emax} - 20.7(2Z_0 - 7) = 0 \quad (6.22)$$

The above quadratic equation produces only one positive root, and this corresponds to the value for R_{Emax}. This is the highest value of R_E for which diode D_L will still be turned on. To make sure we are not creating a marginal situation, we will increase the current through R_E by 30% by choosing a value of R_E that is R_{Emax} divided by 1.3. Table 6.2 shows solutions to (6.22) in the column labeled R_{Emax}, and recommended values of R_E in the next column. These are in very close agreement with those found in figure 5-14 (page 7-53) of [1]. To see how heavily the driving gate is loaded, figure 6.11 was used to calculate I_{OH}. This was then used to find V_{OH} which subtracted from V_{IHmin} (see table 6.1) produced the noise margin V_{NH}. It is noteworthy that the high state noise margin improves (though not very substantially) as Z_0 goes up.

If the need arises to draw a lattice diagram, then an equation for the incident voltage V_i is needed. We can relate V_S to V_{OL} and V_{OH} (for which we already have expressions) by inspection of figure 6.12 to get

$$V_S = V_{OH} + \frac{Z_0}{R_S + Z_0}(V_{OL} - V_{OH}) \quad (6.23)$$

TABLE 6.2 Solutions to (6.22) in the R_{Emax} column. Recommended values of R_E in the next column. Other data of interest are also shown.

Z_0 Ω	R_{Emax} Ω	R_E Ω	I_{OH} mA	V_{OH} mV	V_{NH} mV
50	347	267	13.4	−930	235
62	430	331	10.8	−915	250
75	520	400	9.0	−904	261
90	624	480	7.5	−895	270
100	693	533	6.8	−891	274
120	831	639	5.7	−884	281
150	1038	798	4.5	−877	288

Using (6.4), and observing that $V_{OH} = V(0+)$, we find that

$$V_i = \frac{Z_0}{R_S + Z_0}(V_{OL} - V_{OH}) \tag{6.24}$$

An additional parameter that can be used to verify results obtained in lattice diagrams is the steady-state value of the voltage on the transmission line which can be readily found by inspection from figure 6.11

$$V_{OL(SS)} = \frac{R_E}{R_{OL} + R_E}E_{OL} + \frac{R_{OL}}{R_{OL} + R_E}V_{EE} \tag{6.25}$$

Example 6.4 We wish to see how the values of table 6.2 affect the behavior of series-terminated circuits. Assume that we have 100K ECL working into a 50 Ω transmission line. The model used is that of figure 6.11. The best choice for R_S from (6.17) is 43 Ω. The value for R_E recommended in table 6.2 is 267 Ω.

Before the output changes state we have from (6.18)

$$V_{OH} = -0.930\,\text{V}$$

Since only recommended values are used in this example, we expect the output transistor (represented in our model by the ideal diode D_L) to be forward biased. Accordingly we use (6.19) to find

$$V_{OL} = -1.689\,\text{V}$$

which is below E_{OL}, hence the output transistor will indeed be conducting after switching takes place. From (6.24)

$$V_i = -0.408\,\text{V}$$

From (6.25)

$$V_{OL(SS)} = -1.752\,\text{V}$$

The transmission-line behavior is summarized in figure 6.13.

After $2T$ seconds the returning reflected signal sees an impedance at the source which has a value of $43 + 8||267 = 50.77\,\Omega$ for a very small reflection coefficient of 7.61×10^{-3}. This will eventually cause the voltage in the low state to settle at $V_{OL(SS)}$ within $4T$ seconds.

The signal propagation process is essentially complete after $2T$ seconds. The signal at the load drops to $-1.746\,\text{V}$ which is below the required value of $V_{ILmax} = -1.475\,\text{V}$ after T seconds. The time needed for the input voltage for gates connected anywhere along the middle of the line to drop below V_{ILmax} lies somewhere between T and $2T$ seconds. This is in contrast to the events on a parallel-terminated line where all gates got the proper low level input voltage within T, or less, seconds. ■

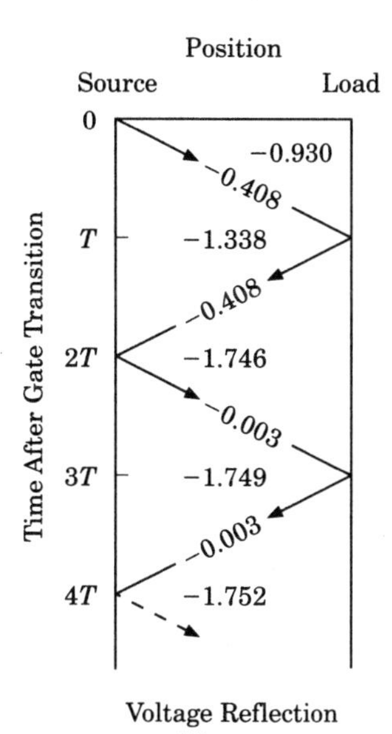

Figure 6.13 Summary of the circuit behavior for 100K ECL operating with a 50 Ω transmission line with $R_S = 43\,\Omega$ and $R_E = 267\,\Omega$.

Example 6.5 As in example 6.4, we have 100K ECL working into a 50 Ω transmission line and R_S is 43 Ω. The value for R_E is $R_{Emax} = 347\,\Omega$.

Before the output changes state we have from (6.18)

$$V_{OH} = -0.912\,\mathrm{V}$$

Since we used for R_E the borderline value from table 6.2, we can use either (6.19) or (6.20) to find

$$V_{OL} = -1.670\,\mathrm{V}$$

and we see that we have a very marginal situation since V_{OL} equals E_{OL}. We will assume that diode D_L is conducting, but we have no assurance that this is indeed the case. From (6.24)

$$V_i = -0.408\,\mathrm{V}$$

From (6.25)

$$V_{OL(SS)} = -1.733\,\mathrm{V}$$

The transmission-line behavior is summarized in figure 6.14. This circuit behaves very much like that in the previous example. The difficulty lies in not knowing if the output transistor of the transmitting gate will be on or off after switching. If it is off, then the above analysis is not correct and has to be redone in a manner similar to that of the next example. ■

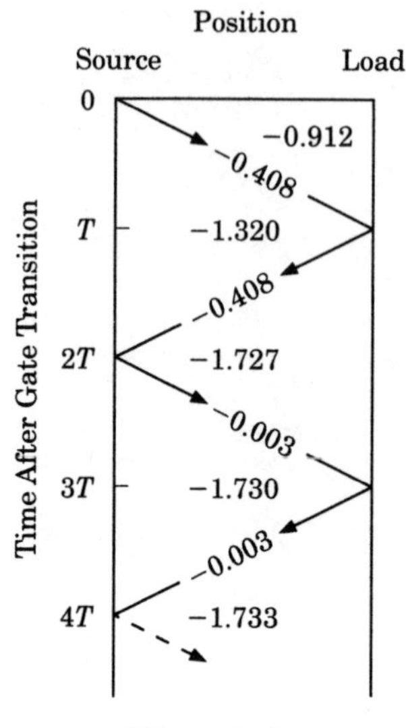

Figure 6.14 Summary of the circuit behavior for 100K ECL operating with a 50 Ω transmission line with $R_S = 43\ \Omega$ and $R_E = R_{Emax} = 347\ \Omega$.

Example 6.6 As in example 6.4 we have 100K ECL working into a 50 Ω transmission line and R_S is 43 Ω. It is desired to apply the theory to a case where R_E is deliberately made too large at a value of 400 Ω.

Before the output changes state we have from (6.18)

$$V_{OH} = -0.904\,\mathrm{V}$$

Since R_E is well above R_{Emax}, we use (6.20) to find

$$V_{OL} = -1.582\,\mathrm{V}$$

and it is very clear that the output transistor is initially cut off. From (6.24)

$$V_i = -0.365\,\mathrm{V}$$

From (6.25)

$$V_{OL(SS)} = -1.725\,\mathrm{V}$$

The transmission-line behavior can be observed in figure 6.15. From the lattice diagram we see that the voltage V_S is equal to −1.924 V an instant after $2T$ seconds, and we assume that diode D_L in figure 6.12 is still turned off. With E_{OL} in figure 6.12 having at this time no influence on the circuit, the voltage

V_S and the voltage V_{EE} determine the voltage V_{OL}. This is calculated, using superposition, to obtain

$$V_{OL} = \frac{R_S}{R_E + R_S} V_{EE} + \frac{R_E}{R_E + R_S} V_S = -2.174\,\mathrm{V}$$

We see from the above that the output transistor turns on after a short time delay Δ, and an additional wave is then launched. The voltage that is established on the entire line drops after time $2T$ to -1.924 V and after time $3T$ it goes as low as -2.215 V. The last two voltages are well below $V_{ILmin} = -1.810$ V.

The shortcomings of choosing too high a value for the pulldown resitor R_E are now apparent. The output transistor becomes cut off during some part of the signal transmission process, thus complicating the analysis. More serious is the resultant undershoot on the line that drives the gate inputs into saturation, causing a reduction in their speed of operation. We conclude that it is best to choose values of R_E from table 6.2 unless there are important overriding reasons for doing otherwise. ■

6.8 Parallel Fanout with Series-Terminated Lines

It is possible to connect multiple gates in the manner shown in figure 6.16 if certain restrictions are observed. We have seen in example 6.4 that for a high-to-low signal transition, with proper series termination, any gates connected along the middle portion of the transmission line will get a sufficiently low input voltage only after the return of the first reflected wave. If speed is the only consideration then it is best to lump all gates very close to the end of the

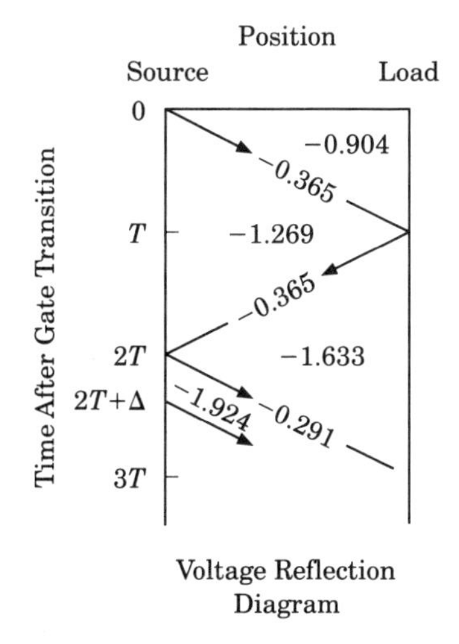

Figure 6.15 Summary of the circuit behavior for 100K ECL operating with a 50 Ω transmission line with $R_S = 43\,\Omega$ and $R_E = 400\,\Omega$.

transmission line, thus avoiding the long delay incurred in waiting for the return wave. But then speed reduction due to capacitive loading, discussed in chapter 3, should be considered.

For 50 Ω transmission lines the pulldown arrangement, which consists of $R_E = 267\,\Omega$ connected to a $-4.5\,\mathrm{V}$ supply, will draw a 13.4 mA current, which still leaves 36.6 mA of the available 50 mA gate output-current for other purposes. In the high state each gate input takes a current I_{IH} which can be as large as 350 μA. This is not likely to be a limitation since many gates loading the transmission line are needed to exceed the 50 mA gate output-current rating. But the problem lies in the fact that the current I_{IH} can cause very significant reductions in the high state noise margin V_{NH}, as will be seen presently.

In example 6.1 we were dealing with an ideal parallel-terminated line. There V_{OH} was $-0.973\,\mathrm{V}$ and it was 192 mV above the voltage V_{IHmin} which is shown in table 6.1. Hence the noise margin V_{NH} in that situation is 192 mV. In example 6.4 it was found that $V_{OH} = -0.930\,\mathrm{V}$, which represents a noise margin of 235 mV. We see that this series-terminated line has initially a 43 mV higher noise margin than the previously mentioned case for parallel termination. Noise reduction due to input gate-current must also be considered.

It is clear from (6.17) and figure 6.11 that for the series-terminated case, the resistance through which the high-state gate-input current is taken is 50 Ω and this reduces the high state noise margin by 17.5 mV for each gate connected as a load. The reduction in noise margin is proportionally more severe for lines with higher characteristic impedances. The 43 mV noise margin advantage for the series-terminated case disappears if we connect more than two gates to the line. If more than 13 gates are connected, then the high state noise margin drops to zero. As a rule of thumb, no more than 10 gates should load a 50 Ω series-terminated transmission-line.

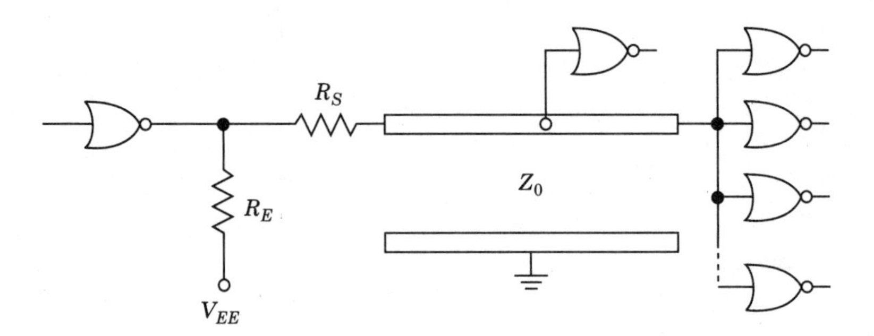

Figure 6.16 A possible connection of gates to a series-terminated line.

The parallel-fanout arrangement shown in figure 6.17 solves two problems. It overcomes some of the loading limitation capability of the single series-terminated line, and it also eliminates the disadvantage of having to lump all loads at one location when using series-terminated lines.

The resistors R_S are chosen to match each transmission line, and (6.17) can be used to find very reasonable values that produce reflection coefficients of only a few percent. We observe in figure 6.17 that the N lines, including their series matching-resistors, are in parallel. The impedance looking into this parallel connection is

$$Z_\pi = \frac{R_S + Z_0}{N} \tag{6.26}$$

To find the value of the pulldown resistor R_E, we simply replace $Z_0 + R_S$ in (6.21) with Z_π to obtain the expression

$$\begin{aligned}(E_{OH} - E_{OL})R_{Emax}^2 - (Z_\pi + R_{OH})(E_{OL} - V_{EE})R_{Emax} \\ - Z_\pi(E_{OL} - V_{EE})R_{OH} = 0\end{aligned} \tag{6.27}$$

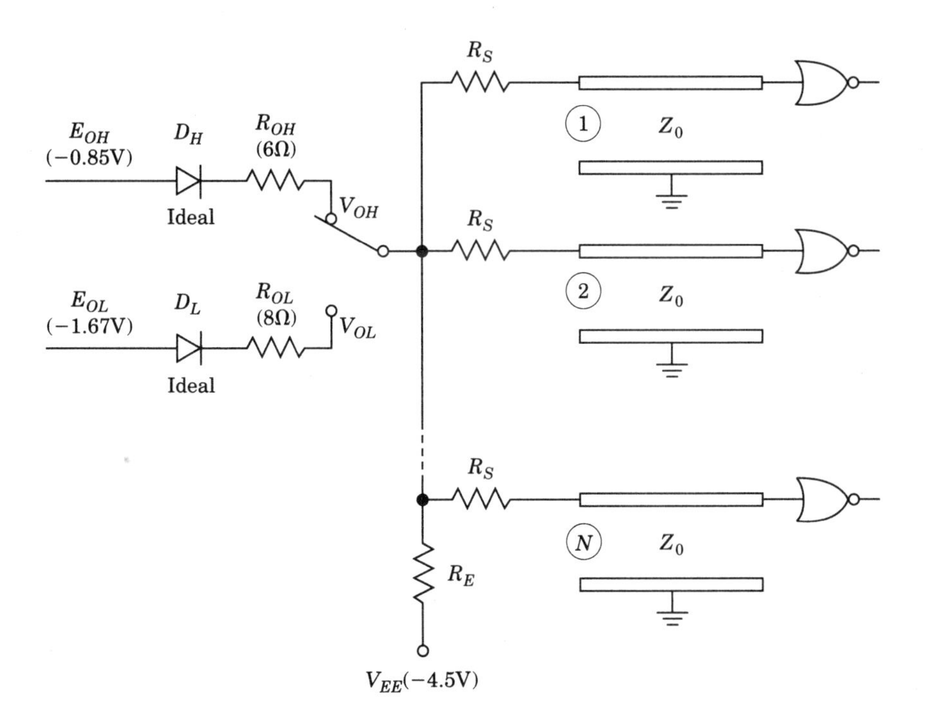

Figure 6.17 Parallel fanout with series-terminated lines.

The above equation can be reduced further by substitution of (6.26) to obtain the form

$$R_{Emax}^2 - 3.45\left(\frac{2Z_0 - 7}{N} + 6\right)R_{Emax} - 20.7\left(\frac{2Z_0 - 7}{N}\right) = 0 \qquad (6.28)$$

For the specific case $Z_0 = 50\,\Omega$ the last equation reduces to

$$R_{Emax}^2 - 3.45\left(\frac{93}{N} + 6\right)R_{Emax} - \frac{1926}{N} = 0 \qquad (6.29)$$

Table 6.3 shows solutions of (6.29) in the R_{Emax} column, and the recommended values of R_E which, as in table 6.1, were obtained by dividing R_{Emax} by a factor of 1.3. The high state output voltage V_{OH} is also given along with the noise margin V_{NH} for each case. The noise margin shown does not take into consideration the high-state gate-input current I_{IH}. In the high state, each of the gates connected to the series-terminated line takes a current I_{IH} whose maximum value is 350 μA. This current causes a steady-state voltage drop across the resistor R_S whose value is $I_{IH}R_S$. If we take the noise margin V_{NH} and divide it by $I_{IH}R_S$, then that is the maximum number of gates that can load each line if the noise margin at each gate is allowed to be zero. How much noise margin one should allow depends on how noisy is the environment of the ECL circuitry. Under favorable circumstances a noise margin of 150 mV is a reasonable value to demand. Accordingly the maximum number of gates per line should be reduced from those found in table 6.3.

The program for obtaining table 6.3 appears at the end of the chapter in listing 6.1. The program was written for use with Microsoft's Quick BASIC. This is a structured BASIC which allows the use of

TABLE 6.3 Solutions to (6.29) for $Z_0 = 50\ \Omega$ appear in the R_{Emax} column. Recommended values of R_E and other parameters of interest are also tabulated.

N	R_{Emax} Ω	R_E Ω	I_{OH} mA	V_{OH} mV	V_{NH} mV	max. number of gates/line
1	347	267	13.4	−930	235	15
2	186	143	24.5	−997	168	11
3	132	102	33.8	−1053	112	7
4	105	81	42.0	−1102	63	4
5	89	69	48.7	−1142	23	1
6	78	60	55.3	−1182	−17	0

subroutine calls by name, and in a manner similar to FORTRAN, allows argument passing in the subroutine call statement.

It can be seen from table 6.3 that no more than five lines should be connected in a parallel fanout to 50 Ω series-terminated lines. When more than five lines are connected, the 50 mA gate output-current rating is exceeded and in addition the noise margin is reduced to zero. For lines with higher characteristic impedance, the values of R_E are correspondingly higher and it is possible to fan out into more lines before the gate output-current limit of 50 mA is reached. Since the values of R_S in that case are higher, fewer gates can be connected to each line before the noise margin drops to zero. It is left to an exercise to extend the table to values of Z_0 other than 50 Ω.

An alternative to the preceding method is the one shown in figure 6.18. A multiple-output buffer/driver is used to drive multiple lines. Series terminations are shown, but any system of termination can be used on any one of the lines without affecting the operation on the others. The various lines shown in 6.18 need not even have identical characteristic impedances.

6.9 Discussion of Parallel and Series Terminations

For large systems where power consumption is an important consideration, it is advantageous to utilize parallel termination to −2 V on all lines. This is the most power-efficient manner of terminating ECL circuitry. The disadvantage is that an additional power supply

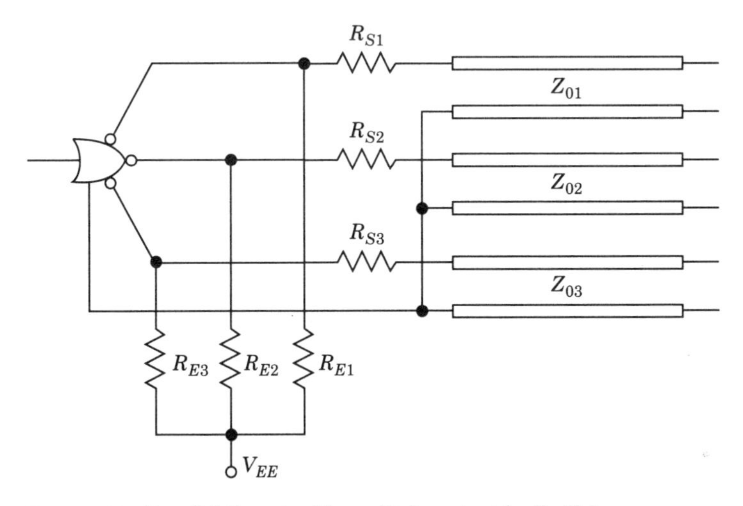

Figure 6.18 Parallel fanout with multiple-output buffer/driver.

and an additional power distribution bus must be provided. Another advantage is that the transmission-line voltage reaches its final value as the first incident wave propagates toward the load end of the line, so that any gates that are connected along the middle portion of the line will get the full input voltage without any undue delay. This is referred to as *incident-wave switching* and it means that gate inputs that are distributed along the length of the line have the full amplitude digital signal delivered to them in the shortest possible time.

The advantage of using the series-termination technique is that only one power supply is required; however, substantially more power is utilized than in the parallel-termination method. The disadvantage is that the transmission-line voltage reaches its final value as the first reflected wave returns to the source end of the line, which is referred to as *reflected-wave switching*. There is, consequently, a substantial delay before the full input voltage reaches any gates that are connected along the middle of the line. It is therefore best to lump all loads at the end of the line which, alas, is accompanied by heavy capacitive loading. A remedy to this is to use the parallel-fanout termination-technique of figure 6.18.

6.10 Unterminated Lines

All digital waveforms require non-zero time intervals to make their voltage transitions. The model that is used to analyze this behavior is the unit step function with ramp transition which was defined in chapter 3. It is reproduced here for convenience

$$u_r^a(t) \equiv \begin{cases} 0, & \text{for } t < 0 \\ t/a, & \text{for } 0 \le t < a \\ 1, & \text{for } a \le t \end{cases} \tag{6.30}$$

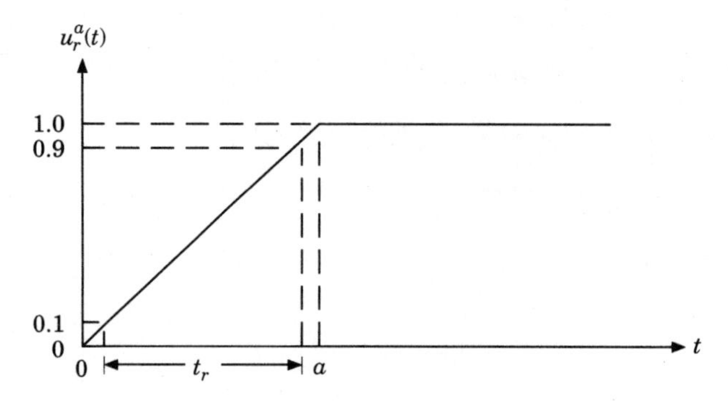

Figure 6.19 Unit step function with ramp transition.

The parameter a represents the *ramp transition time*. It must be reiterated that this parameter must be distinguished from the signal *rise-time* t_r, which is the time required for the signal to go from 10% to 90% of its final value. All of the above parameters are illustrated in figure 6.19.

Transmission lines can be used without the benefit of series or parallel terminations provided the two-way propagation delay on the line is short compared to the signal's ramp transition time. More concisely, the condition is

$$2T \ll a \tag{6.31}$$

and then for all practical purposes the circuitry will not be adversely affected by the reflections. The reasons for this will become clear after the two examples that follow.

Example 6.7 A 50 Ω transmission line has a length of 0.275 ft (8.4 cm) and a propagation delay of 2 ns/ft (0.0656 ns/cm). An 800 mV voltage step with a ramp transition time of 1 ns is applied at $t = 0$ to the left side of the line through a 7 Ω source resistance. We wish to find the voltage at the open circuited receiving end of the transmission line.

The reflection coefficients at the source and load ends are, respectively,

$$\rho_S = -0.754$$

$$\rho_L = 1$$

From the data we also conclude that the one-way propagation delay of this length of line is

$$T = 0.55\,\text{ns}$$

The first incident voltage wave at the sending end is

$$v_{i1}(t) = \frac{Z_0}{R_S + Z_0} u_r^a(t) = 702\, u_r^1(t)$$

and it arrives at the load after a delay of 0.55 ns. Since it encounters an open circuit at the load, it doubles in value upon arrival. The returning wave is reflected at the source with a reflection coefficient of -0.754, and arrives at the load end after a delay of 3 times 0.55 ns. We see that $v_L(t)$, the total voltage at the load, is given by

$$\begin{aligned} v_L(t) = 1404[&u_r^1(t - 0.55) + (-0.754)u_r^1(t - 1.65) \\ &+ (-0.754)^2 u_r^1(t - 2.75) + \cdots] \end{aligned} \tag{6.32}$$

This waveform is shown figure 6.20a. In this case the round-trip delay of the transmission line of 1.1 ns is greater than the ramp transition time of the applied step which is 1 ns. The voltage at the load exhibits such wild oscillations due to the high reflection coefficients at both ends of the line and because each reflected voltage has the time to rise to its final value prior to the arrival of the next reflection. ■

Example 6.8 We have the same data as in example 6.7 except that the line length is only 0.05 ft (0.6 in or 1.52 cm). Now the one-way propagation delay is 0.1 ns. With this choice of line length, (6.31) is reasonably satisfied. The equation for the voltage at the receiving end of the line is now given by

$$v_L(t) = 1404[u_r^1(t - 0.1) + (-0.754)u_r^1(t - 0.3) + (-0.754)^2 u_r^1(t - 0.5) + \cdots] \quad (6.33)$$

This waveform is shown figure 6.20b. In this case the round-trip delay of the transmission line of 0.2 ns is much smaller than the ramp transition time of the applied step which is 1 ns. The voltage at the load is quite well behaved because the oscillations due to the large reflection coefficients have only short time durations to grow in size before another reflection of opposite polarity arrives and causes the voltage to start decreasing. ■

It is clear that as long as (6.31) is satisfied, terminations are not needed for the proper interconnection of digital circuits. The oscillations in the received waveforms are so small that the operation of the digital circuits is not impaired. But it must be kept in mind that

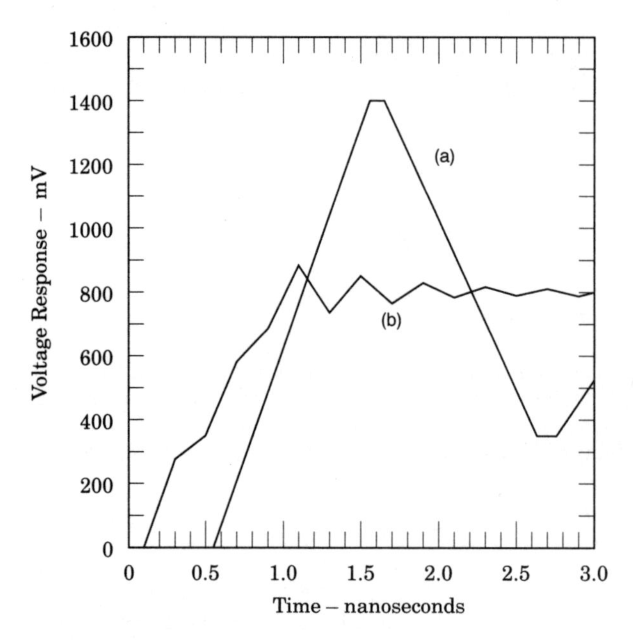

Figure 6.20 Voltage waveforms at the receiving end for (a) example 6.7 and (b) example 6.8.

pulldown resistors are still required for the proper operation of the ECL output circuits. These are needed so that the ECL gate output-transistors will remain in conduction during output transitions from the high to the low state, as pointed out in appendix A. We have seen in examples 6.3 and 6.6 that allowing the output transistor of the sending gate to turn off after a high-to-low output signal transition can cause a substantial undershoot on the line that drives the inputs of the receiving gates into saturation, causing a reduction in their speed of operation.

The placement of the pulldown resistor for an unterminated line is shown in figure 6.21. Comparing this configuration to figure 6.10, we conclude that we need a table of recommended pulldown resistor values similar to that shown in table 6.2. This can be obtained from (6.21) by the simple expedient of setting $R_S = 0$, with the result

$$(E_{OH} - E_{OL})R_{Emax}^2 - (Z_0 + R_{OH})(E_{OL} - V_{EE})R_{Emax} - Z_0(E_{OL} - V_{EE})R_{OH} = 0 \qquad (6.34)$$

For the typical values for 100K ECL shown in figure 6.11, (6.34) reduces to

$$R_{Emax}^2 - 3.45(Z_0 + 6)R_{Emax} - 20.7Z_0 = 0 \qquad (6.35)$$

The above quadratic equation produces only one positive root, and this solution appears in the column labeled R_{Emax} in table 6.4. To avoid a marginal situation, we increase the current through R_E by 30% by choosing a value of R_E that is R_{Emax} divided by 1.3. Table 6.4 shows recommended values of pulldown resistors in the column labeled R_E. As was done in connection with table 6.2, the values of the driving gate-current I_{OH}, the high state output voltage V_{OH} and the noise margin V_{NH} are also tabulated in table 6.4.

It is noteworthy that the pulldown-resistor values shown in table 6.4 for unterminated lines are in all cases smaller than those in table 6.2

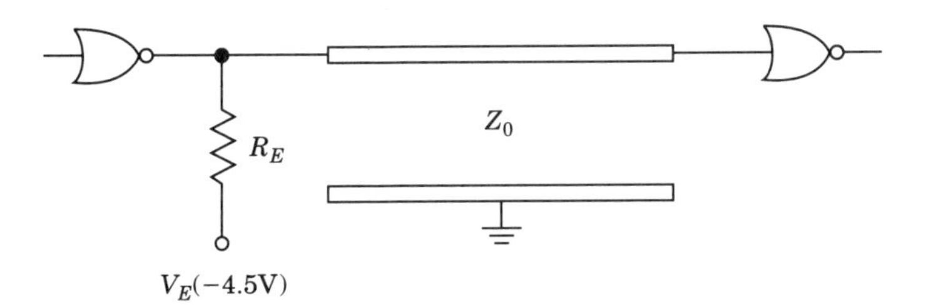

Figure 6.21 Placement of the pulldown resistor for unterminated ECL.

TABLE 6.4 Solutions to (6.35) in the R_{Emax} column. Recommended values of R_E in the next column. Other data of interest are also shown.

Z_0 Ω	R_{Emax} Ω	R_E Ω	I_{OH} mA	V_{OH} mV	V_{NH} mV
50	198	153	23.0	−988	177
62	240	185	19.1	−965	200
75	285	219	16.2	−947	218
90	337	259	13.8	−933	232
100	371	286	12.5	−925	240
120	440	339	10.6	−913	252
150	544	419	8.6	−902	263

for series-terminated lines. The current I_{OH} that the output gate has to deliver in the high output state is higher and the noise immunity is consequently lower for this type of operation when compared to the series-terminated case. In spite of the disadvantages to using unterminated lines, this kind of operation is very convenient because less printed circuit board space is utilized on account of the absence of terminating resistors.

6.11 Conclusion

In this chapter we demonstrated the application of the principles of the earlier chapters to solve some interconnection problems that arise when dealing with fast logic devices. The analysis covered parallel and series termination. It was also pointed out that for short lines satisfying (6.31) it is possible to leave the lines unterminated. The above was done for the 100K series of ECL because these devices are among the fastest logic circuitry commonly available at this time, and have the special requirement of needing pulldown resistors for proper operation. There is no reason why the methods of analysis of this chapter cannot be readily extended to the faster gallium-arsenide devices which are growing in popularity.

References

1. *F100K ECL Logic Databook and Design Guide*, 1990 Edition, National Semiconductor Corporation, Santa Clara, California.
2. W. R. Blood, Jr., *MECL System Design Handbook*, Fourth Edition, Motorola Semiconductor Products Inc., Phoenix, Arizona, 1988.

Program Listing

Listing 6.1: Program written for Microsoft Basic for generating table 6.3.

```
DECLARE SUB root (a!, b!, c!, r1!, r2!)
'Program for generating table 6.3.
'The program below runs on Microsoft Quick BASIC. With
'slight modifications it should run on other BASICs.


'This is the main program
OPEN "table63.dat" FOR OUTPUT AS #1
 ZO = 50
 a = 1
 PRINT #1, "           ZO ="; ZO
 PRINT #1, " N  Remax  RE  IOH  VOH  VNH  Mgates"
 uppr = 2 * x + 4'A varying upper bound on N
 FOR n = 1 TO 6
  b = -3.45 * (6 + (2 * ZO - 7)/n)
  c = -20.71 * (2 * ZO - 7)/n
  CALL root (a, b, c, root1, root2)
  resist = CINT(root1 / 1.3)
  IOH = 1000 * (4.5 - .85) / (resist + 6)
  VOH = -.85 - 6 * (IOH / 1000)
  VNH = 1000 * (VOH + 1.165)
  mgates = INT(VNH / (Z0 - 7) / .35)
  IF (mgates < 0) THEN mgates = 0
  PRINT #1, n; "  ";
  PRINT #1, USING "####  ####  ###.# "; root1, resist, IOH;
  PRINT #1, USING "##.###  ####  ## "; VOH, VNH, mgates
 NEXT n
 PRINT #1,
END


' Subprogram for finding the roots r1, r2 of the
' quadratic equation a*x^2+b*x+c=0
SUB root (a, b, c, r1, r2) STATIC
 disc = SQR(b * b - 4 * a * c)
 r1 = (-b + disc) / (2 * a)
 r2 = (-b - disc) / (2 * a)
END SUB
```

Problems

P6.1 The work leading to the derivation of (6.13) was not presented.

(a) Verify that (6.13) follows from (6.11).

(b) Verify that (6.13) follows (more easily) from (6.12).

P6.2 Is there a reason that (6.13) and (6.21) were derived by considering the output gate high-low transition and not the low-high transition?

P6.3 Verify the calculations and diagrams in examples 6.1, 6.2, and 6.3.

P6.4 In examples 6.1 and 6.2 (6.11) was used to verify that ideal diode D_L was forward biased following the output gate-transition. Show that this conclusion would not have been contradicted had (6.12) been used by mistake in these cases.

P6.5 In example 6.3, (6.12) was used to verify that ideal diode D_L was reverse biased following the gate output transition. Show that this conclusion would not have been contradicted had (6.11) been used by mistake in this case.

P6.6 Repeat the work of examples 6.1, 6.2, and 6.3 for a 75 Ω line with a parallel termination of:

(a) 75 Ω.

(b) 100 Ω.

(c) 120 Ω.

P6.7 Assume that besides having a gate connected at the load we also have gates connected at the sending end and at the center of the line. We are interested in observing when gate switching takes place, the minimum noise margin, and whether there is a risk of gate saturation taking place.

(a) Make your conclusions for example 6.1 by using the lattice diagram of figure 6.6 to draw the voltage at the sending end and at the center of the line.

(b) Make your conclusions for example 6.2 by using the lattice diagram of figure 6.7 to draw the voltage at the sending end and at the center of the line.

(c) Make your conclusions for example 6.3 by using the lattice diagram of figure 6.8 to draw the voltage at the sending end and at the center of the line.

P6.8 For the two circuits shown in figure 6.22 it is desired to compare the power-supply power-requirements. The power-supply power is the power

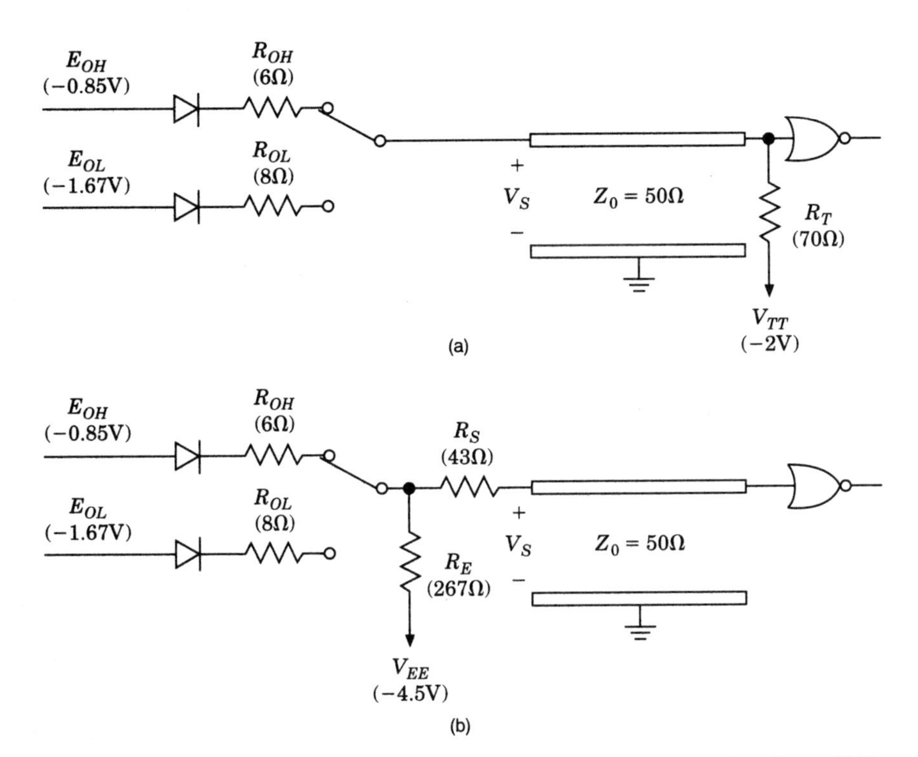

Figure 6.22 Typical circuits for (a) parallel and (b) series termination for a 50 Ω transmission line.

dissipated in all the resistors of the specific circuit. Separately calculate the total power dissipated in the resistors for each of the two connections shown assuming,

(a) The two circuits are in the high-state steady-state.

(b) The two circuits are in the low-state steady-state.

(c) The two circuits are in each of the steady states 50% of the time.

P6.9 Figure 6.22 shows typical circuits for parallel and series termination for a 50 Ω transmission line.

(a) For the two diagrams find the steady-state value for the voltage at the receiving gate when the transmitting gate output is in the high state. What are the noise margins and which termination is superior in this regard?

(b) Repeat part (a) when the transmitting gate is in the low output state.

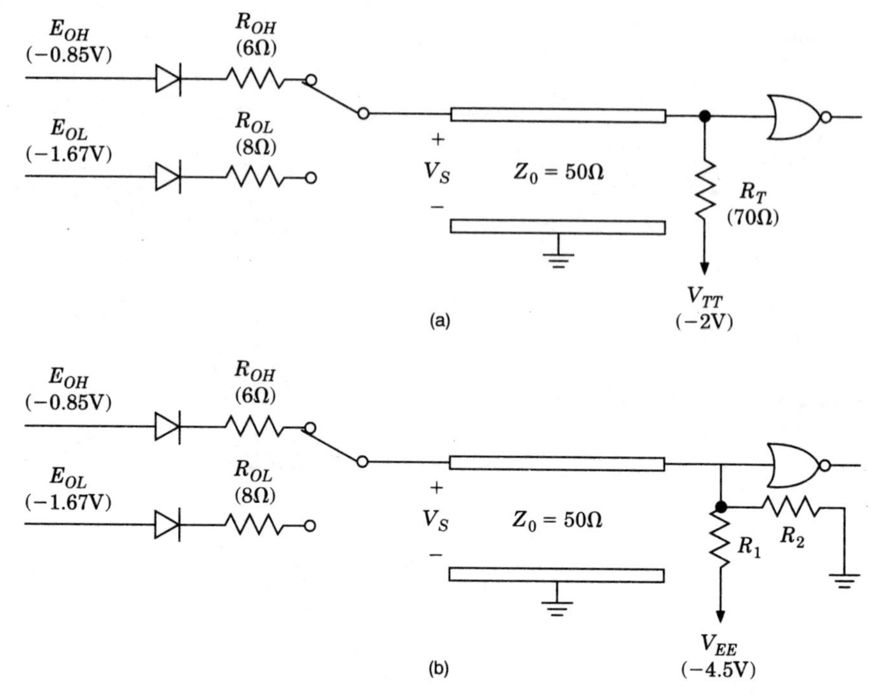

Figure 6.23 Equivalent parallel-termination techniques: (a) with dedicated power supply V_{TT} and (b) with Thevenin equivalent power supply.

(c) The high-state gate-input current is approximately 350 μA. Find the noise margin reduction due to this current for both circuits.

(d) How are the answers in parts (a), (b), and (c) affected when the characteristic impedance of the line is higher? It must be kept in mind that R_S, R_E, and R_T will change.

P6.10 In the parallel-terminated system shown in figure 6.23a it is desired to save the expense associated with the −2 V supply. The Thevenin equivalent form shown in 6.23b will be used instead.

(a) Find the values of R_1 and R_2 for figure 6.23b so that the right-hand termination will be equivalent to the one seen in 6.23a.

(b) For the purpose of this problem ignore transient behavior. Assume each circuit is in the high-state steady-state 50% of the time and in the low-state steady-state the other 50% of the time. Show that the power supply V_{EE} in the circuit shown in 6.23b delivers more power than does the power supply V_{TT} in the circuit shown in 6.23a.

P6.11 Write a computer program for generating the data in table 6.2.

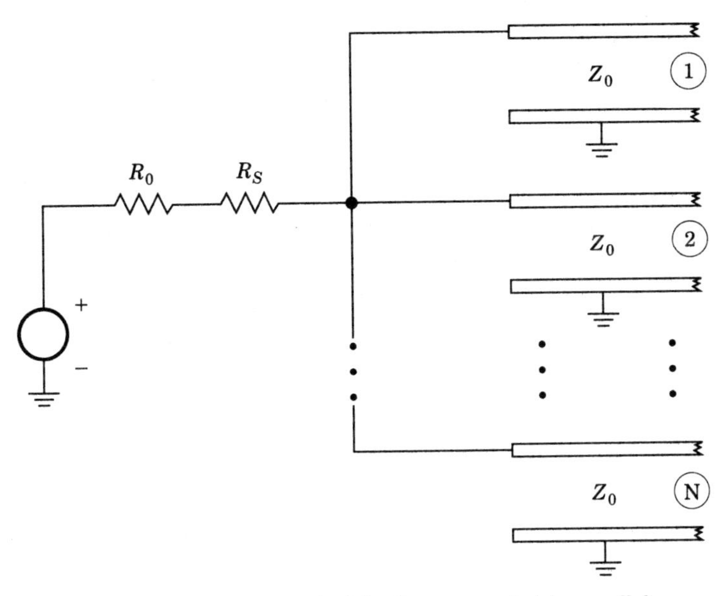

Figure 6.24 A termination method for lines connected in parallel.

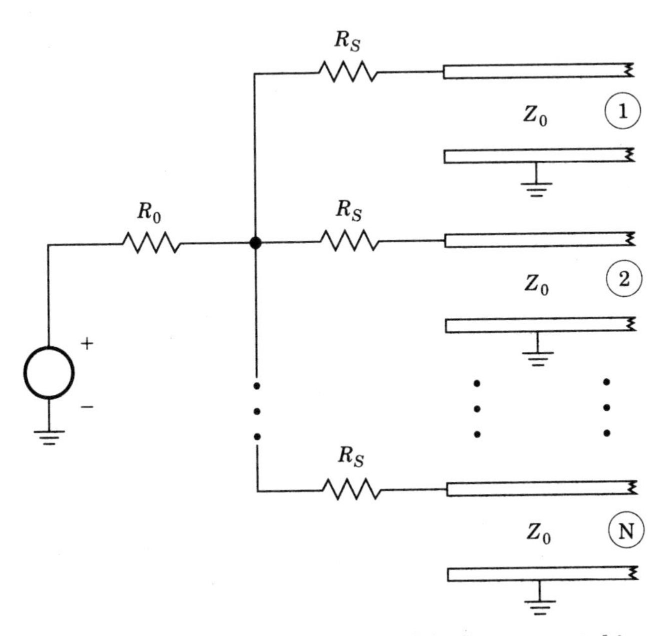

Figure 6.25 Another termination method for lines connected in parallel.

P6.12 It is attempted to match N identical parallel-connected transmission-lines with one resistor R_S, as shown in figure 6.24. Assume the N lines are in parallel and are therefore equivalent to one line.

(a) Find the value of R_S needed to match the lines.

(b) Evaluate the equation for $R_0 = 7\,\Omega$, $Z_0 = 50\,\Omega$ for $1 \le N \le 9$.

(c) Is this method as good as that shown in figure 6.25? (**Hint:** If all else fails, the answer can be found in chapter 2.)

P6.13 It is attempted to match N identical parallel-connected transmission-lines with N resistors of value R_S, as shown in figure 6.25. Assume the N lines (including the resistors R_S) are in parallel and are therefore equivalent to one line.

(a) Find the value of R_S needed to match the lines.

(b) Evaluate the equation for $R_0 = 7\,\Omega$, $Z_0 = 50\,\Omega$ for $1 \le N \le 9$.

(c) Is this method as good as that shown in figure 6.24? (**Hint:** If all else fails, the answer can be found in chapter 2.)

P6.14 Write a program to obtain the values in table 6.3 for the characteristic impedances $Z_0 = 50\,\Omega$, $62\,\Omega$, $75\,\Omega$, $90\,\Omega$, $100\,\Omega$, $120\,\Omega$, and $150\,\Omega$.

Chapter

7

Transmission Line Characteristics

7.1 Introduction

We have found in chapter 1 that the characteristic impedance Z_0 of lossless transmission lines is given by

$$Z_0 = \sqrt{\frac{L_0}{C_0}} \tag{7.1}$$

It was also determined that the speed of propagation is given by

$$\nu = \frac{1}{\sqrt{L_0 C_0}} \tag{7.2}$$

In the above two expressions L_0 is the distributed inductance in henries per meter and C_0 is the distributed capacitance in farads per meter. Since all of the above transmission line parameters are interrelated, then any two parameters can be calculated if any two are given. Thus, for example, if Z_0 and ν are known, then the distributed capacitance and the distributed inductance can be found using

$$C_0 = \frac{1}{\nu Z_0} \tag{7.3}$$

and

$$L_0 = \frac{Z_0}{\nu} \tag{7.4}$$

A knowledge of the distributed capacitance C_0 is essential in order to determine the effect of distributed capacitive-loading on the transmission line. In chapter 3 it was shown that if an additional capacitance C is distributed uniformly along the length l of the line, then the characteristic impedance takes on the modified value

$$Z_{0M} = \frac{Z_0}{\sqrt{1 + C/lC_0}} \tag{7.5}$$

The additional capacitance also affects the speed of propagation

$$\nu_M = \frac{\nu}{\sqrt{1 + C/lC_0}} \tag{7.6}$$

as well as the one-way transmission-line delay

$$T_M = T\sqrt{1 + C/lC_0} \tag{7.7}$$

Example 7.1 A transmission line is available with $Z_0 = 72\,\Omega$ and $\nu = 0.6c$ where c is the speed of light in free space which is 3×10^8 m/s. A 30 cm section is loaded with 12 uniformly spaced gates, each having an input capacitance of 3 pF. We wish to determine Z_{0M}, ν_M, and T_M of this capacitively-loaded line.

We first observe that the speed of propagation of this cable before loading is

$$\nu = 0.6(3 \times 10^8) = 1.8 \times 10^8 \text{ m/s}$$

and the one-way propagation-delay is

$$T = \frac{l}{\nu} = \frac{0.3}{1.8 \times 10^8} = 1.67 \text{ ns}$$

Before proceeding to solve the problem we use (7.3) to determine

$$C_0 = \frac{1}{\nu Z_0} = \frac{1}{(1.8 \times 10^8)72} = 77 \text{ pF/m}$$

We now evaluate

$$\sqrt{1 + C/lC_0} = \sqrt{1 + 36/(0.3 \times 77)} = 1.6$$

which we substitute into (7.5), (7.6), and (7.7) to evaluate the modified characteristic-impedance

$$Z_{0M} = 45\,\Omega$$

the modified speed of propagation

$$\nu_M = 1.13 \times 10^8\,\text{m/s}$$

and the modified one-way delay

$$T_M = 2.67\,\text{ns}$$ ■

The sections that follow present the equations needed to determine the characteristic impedance and the speed of propagation for five transmission-line geometries which are commonly found in computer circuit interconnections. Only the final results are presented, since the strict mathematical derivation of these equations is beyond the scope of this book.

7.2 Coaxial Transmission-Line

The geometry of a coaxial transmission-line is shown in figure 7.1. This line is made up of an inner and outer conductor separated by a dielectric. The line has a circular cross section. This kind of line is well shielded from external interference and is used to connect systems that carry weak signals over relatively long distances.

The characteristic impedance of the coaxial transmission-line can be found on page 252 of [1] and a detailed derivation for this case appears in [2]. It is given by

$$Z_0 = \frac{60}{\sqrt{\epsilon_r}} \ln\left[\frac{D}{d}\right] \tag{7.8}$$

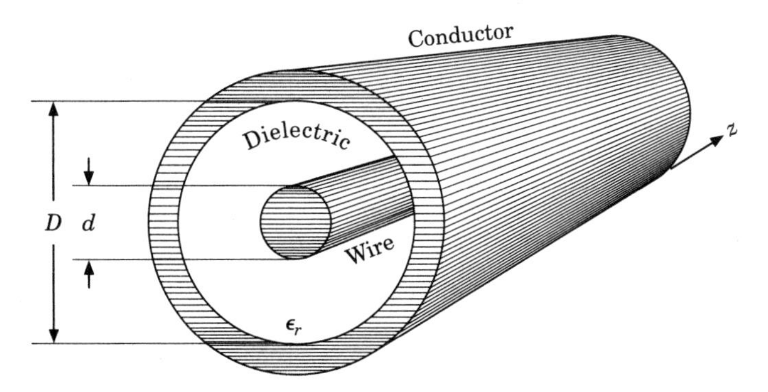

Figure 7.1 Geometry of a coaxial transmission-line.

where it has been assumed that we are dealing with a non-ferromagnetic dielectric separating the inner and outer conductors. This has a relative magnetic-permeability of free space, namely $\mu_r = 1$, and a relative dielectric-permittivity ϵ_r.

The speed of propagation on this line is

$$\nu = \frac{c}{\sqrt{\epsilon_r}} \tag{7.9}$$

Example 7.2 An RG-58/U 50 Ω coax has a polyethylene dielectric. The speed of propagation on this line is (2/3)c. The outer diameter of the dielectric is 3 mm. Find the dielectric constant of the polyethylene, the diameter of the inner conductor, and the distributed capacitance for this cable.

Using (7.9) we find

$$\epsilon_r = \left[\frac{c}{\nu}\right]^2 = \left[\frac{c}{(2/3)c}\right]^2 = 2.25$$

From (7.8) we get

$$d = D\,e^{-Z_0\sqrt{\epsilon_r}/60} = 0.86\,\text{mm}$$

And finally from (7.3) we obtain

$$C_0 = \frac{1}{\nu Z_0} = 100\,\text{pF/m}$$

■

7.3 Parallel Wire and Twisted Pair

The geometry of a parallel-wire transmission-line is shown in figure 7.2. This consists of two parallel conductors separated by a dielectric medium of relative dielectric-constant ϵ_r. This line is very often used to interconnect subassemblies, such as computers and printers, over distances not exceeding a few meters. In that case it takes the form of flat ribbon-cable. Typically, when parallel wire cable is constructed using 26 gauge (0.4 mm) hookup-wire, it produces cable whose Z_0 lies in the 100 Ω to 120 Ω range. The same holds true for twisted pair provided the twists are kept below 30 twists per foot (or one twist per centimeter).

The characteristic impedance of the parallel-wire transmission-line can be found on page 252 of [1] and a detailed derivation for this case can be found in [2]. The equation is*

$$Z_0 = \frac{120}{\sqrt{\epsilon_r}} \cosh^{-1}\left[\frac{S}{d}\right] \tag{7.10}$$

*It may sometimes be more convenient to substitute $\cosh^{-1}(x) = \ln\left[x + \sqrt{x^2 - 1}\right]$.

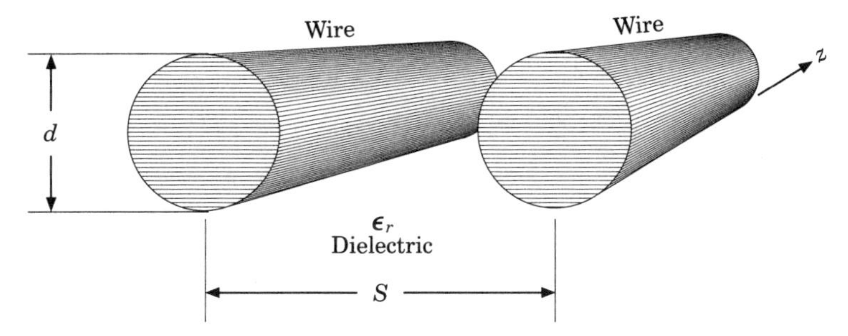

Figure 7.2 Geometry of a parallel-wire transmission-line.

The speed of propagation on this line is determined by the dielectric constant of the medium separating the two conductors, hence (7.9) remains applicable in this case.

Example 7.3 24 gauge hookup wire has a conductor diameter of 0.0201 in. (0.511 mm). The polyethylene insulation ($\epsilon_r = 2.3$) is 0.010 in. (0.254 mm) thick. This is formed into a twisted pair. We wish to determine the characteristic impedance of this hookup.

The outer diameter of the wire is clearly 0.0401 in. (1.019 mm). When twisting the wires the conductors end up with their outer insulation in close contact. Consequently the spacing S between the two wires ends up being the same as the outer diameter of the wire. We can now use (7.10) to find

$$Z_0 = \frac{120}{\sqrt{2.3}} \cosh^{-1}\left[\frac{0.0401}{0.0201}\right] = 104\,\Omega$$ ■

7.4 Wire over Ground Plane

A single wire above a ground plane as shown in figure 7.3 has the characteristics of a transmission line. This often arises when it is necessary to connect points on printed circuit boards with hookup wire and also when using wire-wrap connections. If the distance between the wire and the printed circuit board is not well controlled, then the characteristic impedance varies with distance. This has the same effect as transmission lines connected in cascade. We have seen in chapter 2 that this causes reflections. Proper precautions should be taken when using this kind of wiring.

The equation for the characteristic impedance of this transmission line follows from that for the parallel-wire line. It is obtained by halving the expression given in (7.10), and also replacing the variable S with $2h$. The resultant equation is

$$Z_0 = \frac{60}{\sqrt{\epsilon_r}} \cosh^{-1}\left[\frac{2h}{d}\right] \tag{7.11}$$

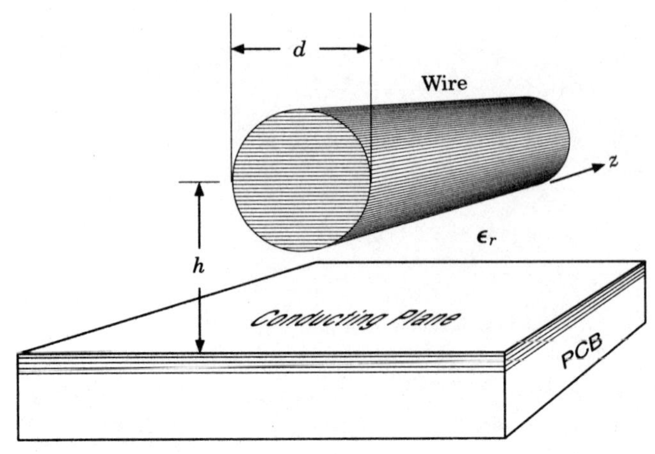

Figure 7.3 Single wire over a ground plane.

The speed of propagation on this line is determined by the dielectric constant of the medium separating the wire from the ground plane and is given by (7.9).

Example 7.4 30 gauge wire-wrap wire has a conductor diameter of 0.0100 in. (0.254 mm). The polyethylene insulation has $\epsilon_r = 2.3$. The diameter including the insulation is 0.019 in. (0.483 mm). This wire is lying in contact with the printed-circuit-board conducting surface. We wish to determine the characteristic impedance of this connection.

The value $2h$ is clearly equal to the outer diameter of the wire. Putting this into (7.11), we obtain

$$Z_0 = \frac{60}{\sqrt{2.3}} \cosh^{-1}\left[\frac{0.019}{0.0100}\right] = 50\,\Omega$$

The question of how the impedance varies as a function of the spacing above the PC board conducting-surface is left as an exercise. ■

7.5 Stripline

As can be seen in figure 7.4, the stripline consists of a central rectangular conductor buried in a dielectric, surrounded by two ground planes. This type of configuration is found in multilayer printed circuit-boards. It is used to increase the component density on a printed circuit-board.

The literature contains many different equations that can be used to evaluate the characteristic impedance of striplines. Alas, the majority assume that $t = 0$. Cohn, in [3], presents (fairly elaborate) equations that produce excellent results, and they include consideration of a non-zero t. In his paper, Cohn took the trouble to show that the results

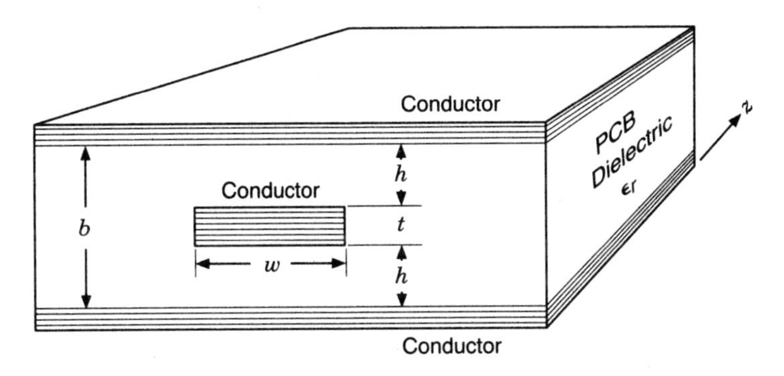

Figure 7.4 Stripline geometry.

produced by the following equations are in excellent agreement with physical data.

The equation for the characteristic impedance is given by

$$Z_0 = \frac{30\pi}{\sqrt{\epsilon_r}[D + \xi w/b]} \tag{7.12}$$

where

$$\xi = \frac{1}{1 - t/b} \tag{7.13}$$

and

$$D = \frac{1}{\pi}[2\xi \ln(1 + \xi) - (\xi - 1)\ln(\xi^2 - 1)] \tag{7.14}$$

If the stripline parameters fall within particular bounds, then the simpler equation that follows, which appears in [4–7], can be used for rapid calculator evaluations. The equation for Z_0 takes the form

$$Z_0 = \frac{60}{\sqrt{\epsilon_r}} \ln\left[\frac{4b}{0.67\pi(0.8w + t)}\right] \tag{7.15}$$

Unlike (7.12) the last equation has a parameter bound on accuracy. The above equation is most accurate provided

$$\frac{w}{b - t} < 0.35 \quad \text{and} \quad \frac{t}{b} < 0.25 \tag{7.16}$$

If the stripline parameters are such that the above bounds are violated, then the results of (7.15) become inaccurate and the use of Cohn's equation is very strongly recommended.

Since the signal is confined entirely to the interior of the dielectric, (7.9) remains applicable for the calculation of signal speed.

Example 7.5 A stripline is fabricated using a G-10 glass-epoxy PC board whose thickness is $b = 0.031$ in. For this material $\epsilon_r = 4.7$. It is further given that $t = 0.0014$ in. and $w = 0.010$ in. It is desired to find the characteristic impedance.

Substitution into (7.13) produces the result $\xi = 1.047$. This is used in (7.14) with the result $D = 0.513$. When the substitution of the above two quantities is made into (7.12), we get the result $Z_0 = 51.1\,\Omega$. On the other hand, (7.15) gives the result $Z_0 = 50.8\,\Omega$, in good agreement with the previous result. It is of no great use to be able to calculate Z_0 to too high an accuracy. In practice, errors in controlling all the given dimensions during fabrication, as well as variations in the value of ϵ_r, can produce departures from calculated answers that are substantially greater than the calculation errors. ■

7.6 Microstrip

As can be seen in figure 7.5, the microstrip is representative of the situation encountered for boards with conductors on only two sides. The transmission line consists of a conductor on top of a dielectric, with a ground plane on the bottom side of the dielectric. The equations presented here were taken from [8]. They have been verified against the table of physical data found in [9] and it was determined that they are very accurate. In contrast to the much older (but simpler) equations found in [9], the equations that follow have no range restriction on them.

The equations for characteristic impedance are given by

$$Z_0 = \frac{60}{\sqrt{\epsilon_{re}}} \ln\left[\frac{8h}{w_e} + \frac{w_e}{4h}\right], \qquad \text{for } \frac{w}{h} \le 1 \tag{7.17}$$

$$Z_0 = \frac{120\pi}{\sqrt{\epsilon_{re}}} \left\{\frac{w_e}{h} + 1.393 + 0.667 \ln\left[\frac{w_e}{h} + 1.444\right]\right\}^{-1}, \qquad \text{for } \frac{w}{h} > 1 \tag{7.18}$$

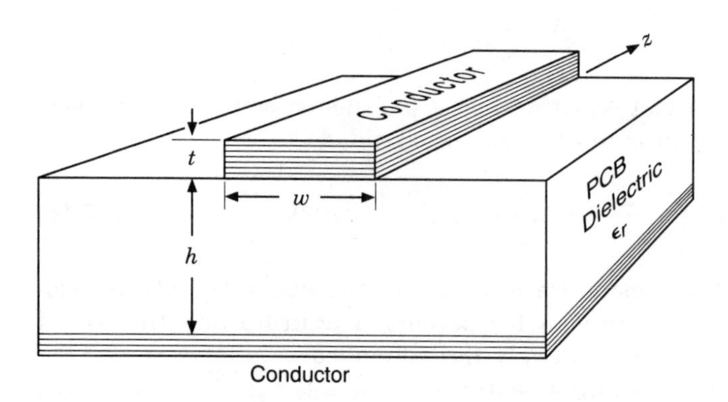

Figure 7.5 Microstrip geometry.

where the effective width is given by

$$\frac{w_e}{h} = \frac{w}{h} + \frac{1.25}{\pi}\frac{t}{h}\left[1 + \ln\left(\frac{4\pi w}{t}\right)\right], \qquad \text{for } \frac{w}{h} \leq \frac{1}{2\pi} \tag{7.19}$$

$$\frac{w_e}{h} = \frac{w}{h} + \frac{1.25}{\pi}\frac{t}{h}\left[1 + \ln\left(\frac{2h}{t}\right)\right], \qquad \text{for } \frac{w}{h} > \frac{1}{2\pi} \tag{7.20}$$

and

$$\epsilon_{re} = \frac{\epsilon_r + 1}{2} + \frac{\epsilon_r - 1}{2\sqrt{1 + 10h/w}} - \frac{(\epsilon_r - 1)}{4.6}\frac{t/h}{\sqrt{w/h}} \tag{7.21}$$

The use of an effective width w_e and that of an effective relative dielectric-constant ϵ_{re} accounts for the fact that the signal propagates partly in the dielectric and partly in the air above. As a consequence it is necessary to adjust the parameters needed to perform the calculations. The constant ϵ_{re} can be used to find the speed of propagation on this line by using

$$\nu = \frac{c}{\sqrt{\epsilon_{re}}} \tag{7.22}$$

The above equations have the advantage that they are accurate irrespective of the parameters of the microstrip. There are a simpler (and older) set of equations due to Kaupp [9]. They have some shortcomings. The equation for Z_0 produces very accurate results for a restricted range of parameters. The equation for ϵ_{re} is entirely independent of the microstrip geometry, so the results it produces in (7.22) are approximate.

The equation for Z_0 is given by

$$Z_0 = \frac{60}{\sqrt{\epsilon_{re}}}\ln\left[\frac{4h}{0.67(0.8w + t)}\right] \tag{7.23}$$

where the relative effective dielectric-constant is given by

$$\epsilon_{re} = 0.475\epsilon_r + 0.67 \tag{7.24}$$

The equation for Z_0 produces very accurate results for the range of parameters satisfying

$$\frac{w}{h} < 1.25, \qquad 0.1 < \frac{t}{w} < 0.8, \qquad 2.5 < \epsilon_r < 6 \tag{7.25}$$

As for the first set of equations, the result of (7.24) can be substituted into (7.22) to find the speed of propagation on the microstrip line.

Example 7.6 A microstrip line is fabricated using a $\frac{1}{32}$ in. G-10 glass-epoxy PC board whose $\epsilon_r = 4.7$. From the specification $h = 0.032$ in. It is further given that $t = 0.0014$ in. and $w = 0.010$ in. It is desired to find the characteristic impedance and the speed of propagation on the line and to compare the results produced by the equations taken from [8] and [9].

Substitution into (7.23) produces the result $Z_0 = 106\,\Omega$. In addition we find using (7.24) that $\epsilon_{re} = 2.90$ which is used in (7.22) to find that $\nu = 0.587c$. On the other hand, when the more accurate (7.17–7.20) are used, it is found that $Z_0 = 102.4\,\Omega$. From (7.21) it is found that $\epsilon_{re} = 3.11$ which when substituted into (7.22) produces $\nu = 0.567c$.

In this case it hardly made a difference which equations were used to solve the problem. But for certain choices of parameters there can be large differences in the results produced by the two sets of equations. ■

7.7 The Parallel-Plate Transmission-Line

The parallel-plate transmission-line is used in multilayer PC boards to distribute power to the various integrated circuits (ICs). Paths such as stripline and microstrip produce characteristic impedances that are quite high. Consider a power supply connected to an IC by means of a $50\,\Omega$ transmission line. Suppose the IC goes from an idle state, where it draws almost no current from the power supply, to an active state where it suddenly requires a current of 20 mA. In the absence of surge suppression capacitors, this current represents a voltage dip of $20\,\text{mA} \times 50\,\Omega = 1\,\text{V}$. If the characteristic impedance of the connecting line were less than $1\,\Omega$, then the voltage dip would not be very severe.

The configuration we are dealing with is that shown in figure 7.6. This is a parallel plate line of width w and conductor separation b. The characteristic impedance of this line is readily found [1] and is given by

$$Z_0 = \frac{120\pi}{\sqrt{\epsilon_r}} \cdot \frac{b}{w} \qquad \text{for } b \ll w \tag{7.26}$$

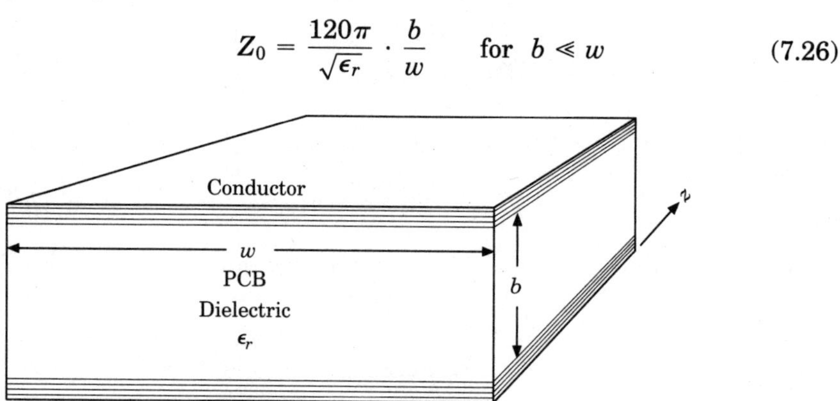

Figure 7.6 Parallel-plate transmission-line geometry.

Example 7.7 In a multilayer PC board one conducting plane is used to supply +5 V and an adjacent plane is used as a ground. This plane is 100 mm (4 in.) wide and the glass-epoxy ($\epsilon_r = 4.7$) dielectric-thickness is 0.4 mm ($\frac{1}{64}$ in.). Find the characteristic impedance and determine the current transient required to cause a voltage change of 0.25 V.

Substitution into (7.26) gives the result $Z_0 = 0.7\,\Omega$. A chip would have to draw a transient current of 360 mA to make the voltage dip a relatively insignificant 0.25 V. It is now apparent why it is good practice in multilayer PC boards to dedicate two whole conducting planes to the distribution of power. ■

7.8 Conclusion

A book on the subject of transmission lines would not have been complete without a chapter useful for determining transmission line parameters. This chapter serves that purpose. In all cases an effort was made to present the most accurate equations to be found in the literature. One is tempted at times to present more intricate equations, but these have to be discarded when it is found that they do not produce results that are in agreement with practical reference data.

The information found in this chapter can be useful for determining the characteristic impedance and the speed of propagation on a transmission line whose geometry is known. It is also useful for solving the inverted problem, that of choosing a geometry in order to obtain a desired performance parameter. At times the line geometry may not fit exactly one or another of the cases described, in which case the equations can be used as an estimate to assess the system performance.

References

1. S. Ramo, J. R. Whinnery, and T. Van Duzer, *Fields and Waves in Communication Electronics,* Second Edition, John Wiley & Sons, New York, 1984.
2. D. K. Cheng, *Field and Wave Electromagnetics,* Second Edition, Addison-Wesley, Reading, Massachusetts, 1989.
3. S. B. Cohn, "Problems in Strip Transmission Lines," *IRE Transactions on Microwave Theory and Technique,* Vol. MTT-3, pp. 119-126, March 1955.
4. *FAST Logic Applications Handbook,* 1990 Edition, National Semiconductor Corporation, Santa Clara, California.
5. *FACT Advanced CMOS Logic Databook,* 1990 Edition, National Semiconductor Corporation, Santa Clara, California.
6. *F100K ECL Logic Databook and Design Guide,* 1990 Edition, National Semiconductor Corporation, Santa Clara, California.
7. W. R. Blood, Jr., *MECL System Design Handbook,* Fourth Edition, Motorola Semiconductor Products Inc., Phoenix, Arizona, 1988.
8. K. C. Gupta, R. Garg, and R. Chadha, *Computer-Aided Design of Microwave Circuits,* Artech House, Dedham, Massachusetts, 1981, pp. 61–62.
9. H. R. Kaupp, "Characteristics of Microstrip Transmission Lines," *IEEE Transactions on Electronic Computers,* Vol. EC-16, April 1967.

Problems

P7.1 A 20 cm length of 90 Ω line is loaded with 8 uniformly spaced gates each having an input capacitance of 4 pF. The speed of propagation on the unloaded line is $\nu = 0.7c$. Determine Z_{0M}, ν_M, and T_M of this capacitively-loaded line.

P7.2 An RG-59/U 75 Ω coax has a polyethylene dielectric. The inner conductor has a diameter 0.023 in. and the inner diameter of the outer conductor is 0.146 in. The polyethylene dielectric has a relative dielectric-constant $\epsilon_r = 2.3$. Find the characteristic impedance and the speed of propagation on this cable. Also determine the distributed capacitance and inductance per meter.

P7.3 The conductor diameter of 26 gauge hookup wire is 0.0159 in. (0.404 mm). The polyethylene insulation ($\epsilon_r = 2.3$) is 0.010 in. (0.254 mm) thick. This is formed into a twisted pair. Find the characteristic impedance of this configuration.

P7.4 Assume the same wire as in example 7.4. To see how the characteristic impedance varies as a function of the spacing above the PC board conducting-surface, assume that the wire is now suspended one wire diameter above the PC board.

(a) Determine the characteristic impedance of this wire.

(b) Determine the characteristic impedance of this wire on the assumption that $\epsilon_r = 1$.

(c) It can be argued that when the wire is that high above the board some of the electromagnetic signal travels inside the insulation and some in the air surrounding the wire. It is then reasonable to say that the actual characteristic impedance lies somewhere between the two solutions obtained above. One way to deal with this problem might be to take the geometric mean of the two solutions.

(d) Compare the Z_0 obtained above with that found in example 7.4. If the wire cannot be kept in intimate contact with the PC board, then are large reflections to be expected? What is the reflection coefficient?

P7.5 As in example 7.5, a stripline is fabricated using a $\frac{1}{16}$ in. G-10 glass-epoxy PC board whose $\epsilon_r = 4.7$. The other parameters are given that $t = 0.0014$ in. and $w = 0.010$ in. Find the characteristic impedance by the two equations given and compare. Are any of the bounds given in connection with (7.15) violated?

P7.6 A microstrip line is fabricated using a $\frac{1}{16}$ in. G-10 glass-epoxy PC board whose $\epsilon_r = 4.7$. It is further given that $t = 0.0028$ in. and $w = 0.024$ in. It is desired to find the characteristic impedance and the speed of propagation on the line.

P7.7 For the specifications of the previous problem, find the trace width w needed to produce a line with $Z_0 = 50\,\Omega$.

P7.8 For the specifications given in example 7.6, determine the trace width w needed to produce a line with $Z_0 = 50\,\Omega$.

Appendix

A

Introduction to High Speed ECL

A.1 Introduction

Among the fastest of the well-established digital building block circuits available at the present time are the 100K series of emitter coupled logic, commonly referred to as ECL. The propagation delay of a signal through some types of 100K gates is less than 1 ns. We will not go into all the details of device operation, but we need a good idea of the behavior at the input and output terminals, since this will influence our analysis of the signal propagation along the circuit interconnections.

The designation 100K means that this series of digital logic building blocks carries device numbers in the 100000 range. For example, the integrated circuit whose device number is 100121 consists of 9 inverters in one 24 pin package. Other devices within the 100K designation are available to fill various other requirements in digital circuit design.

A fictitious ECL gate is shown in symbolic form in figure A.1. If such a gate existed, it would probably be called an Inverting/Non-inverting Buffer. This gate is shown in general schematic form in figure A.2.

$V_{01} = V_I$

V_I

$V_{02} = \overline{V_I}$

Figure A.1 Symbolic form of an ECL gate.

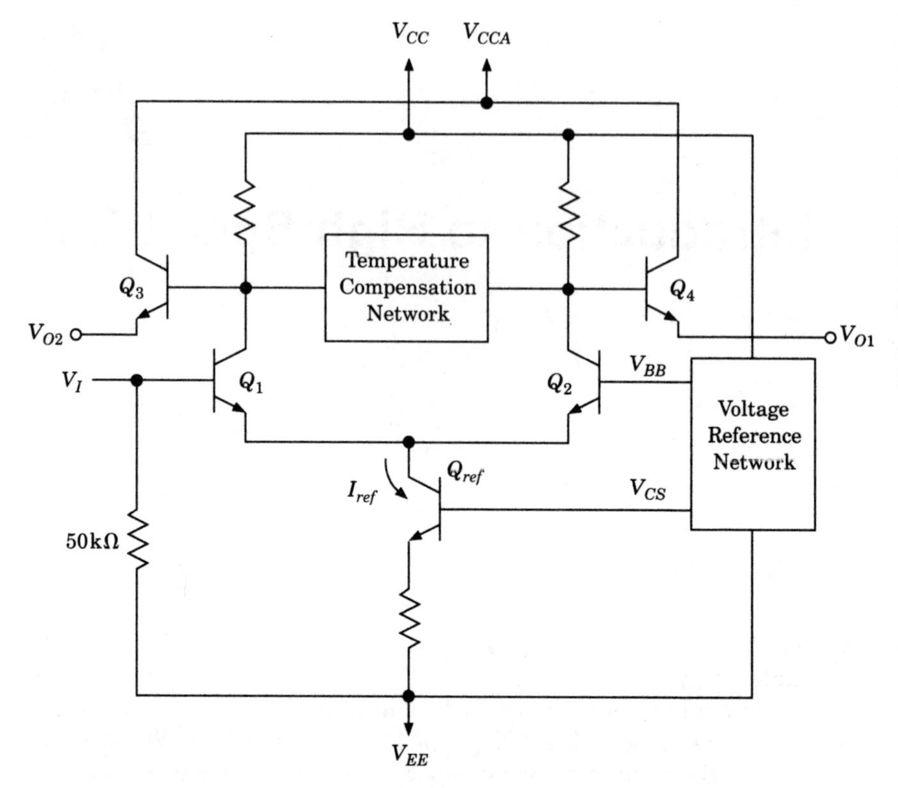

Figure A.2 General schematic form of an ECL gate.

The voltage reference network is temperature compensated to supply the constant voltage references V_{BB} and V_{CS} [1,2]. The constant reference voltage V_{CS} makes transistor Q_{ref} draw a constant current I_{ref} at its collector. This means that the sum of the emitter currents of transistors Q_1 and Q_2 is a constant determined by the collector current of Q_{ref}. The reference voltage V_{BB} has a mean value of -1.32 volts, and there is a ± 40 mV manufacturing lot-to-lot variation. If the input voltage V_I equals V_{BB}, then the emitter currents of the two transistors Q_1 and Q_2 will be equal. If the input voltage V_I rises above V_{BB} by 125mV, then transistor Q_1 takes essentially all of the current I_{ref}. Under those circumstances the voltage at collector Q_2 rises to the voltage V_{CC} (usually ground), while the voltage at collector Q_1 drops to approximately -0.9 V. Conversely if the input voltage V_I drops below V_{BB} by 125 mV, then transistor Q_2 takes essentially all of the current I_{ref}. Under those circumstances the voltage at collector Q_1 rises to the voltage V_{CC} (usually ground), while the voltage at collector Q_2 decreases to -0.9 V. Although an input signal swing of 250 mV is sufficient to effect a change of state at the gate output, a figure of 750 mV is usually specified to provide for noise immunity and to allow for the lot-to-lot variation in the voltage V_{BB}.

The above input voltage swings are useful numbers, but there are instances where more detailed specifications are needed. In that case we make reference to the high state input voltage V_{IH}, which can be as low as $V_{IHmin} = -1.165\,\text{V}$ and still cause a transition to the high state in all manufactured units. ECL derives some of its speed from the fact that the transistors that are used in the gate are not allowed to go into saturation. It is therefore important that the input voltage V_I not exceed $V_{IHmax} = -0.880\,\text{V}$, otherwise saturation of transistor Q_1 will take place. We also refer to the voltage V_{IL}, which can be as high as $V_{ILmax} = -1.475\,\text{V}$ and still cause a transition to the low state in all manufactured units. It is important that the input voltage V_I not fall below $V_{ILmin} = -1.810\,\text{V}$, otherwise saturation of transistor Q_2 will take place, slowing the operation of the ECL gate. Table A.1 summarizes the above specifications.

ECL gates are in many instances constructed to produce complementary output signals, such as those that appear at the collectors of Q_1 and Q_2 in figure A.2. These signals cannot be used as inputs to other ECL gates since they do not correspond to proper ECL voltage logic levels. Also, the output impedances at the collectors of transistors Q_1 and Q_2 are too high to drive devices whose input impedance might be in the range of 50 Ω. As a consequence the emitter follower transistors Q_3 and Q_4 are used to shift the output signals to the proper ECL logic levels and to lower the gate output impedance to approximately 7 Ω. The emitter outputs V_{O1} and V_{O2} have no internal resistors connected to them. It will be seen later that external pulldown resistors are required for the proper operation at the gate output.

The output voltages V_{O1} and V_{O2} are compensated by the temperature compensation network shown in figure A.2 to prevent any significant variation in the 0°C to +85°C temperature range. For

TABLE A.1 Some Critical Values for the Input voltage V_I for 100K ECL.

V_I	Gate input-voltage
$V_{IHmax} = -0.880$ V	Saturation of an input transistor risked above this value of V_I.
$V_{IHmin} = -1.165$ V	All manufactured units will make a transition to the high state when the input voltage exceeds this value.
$V_{ILmax} = -1.475$ V	All manufactured units will make a transition to the low state when the input voltage falls below this value.
$V_{ILmin} = -1.810$ V	Saturation of an input transistor risked below this value of V_I.

the output logic swing, a figure of 800 mV is mentioned in casual reference to 100K ECL. More specific numbers can be obtained from the manufacturer's specifications. Although the output circuitry of ECL is somewhat nonlinear, it is adequate for our purposes to use the approximate linearized model, as shown in figure A.3.

The ideal diodes D_H and D_L along with the output resistances R_{OH} and R_{OL} represent the effect of the emitter follower output driver of the ECL gate in the high and low output voltage state, respectively. The ideal diodes are needed to account for the fact that the emitter follower output transistors Q_3 and Q_4 (in figure A.2) will not allow current conduction into the emitter. In this model, the diodes are ideal, namely D_H behaves like a short circuit as long as the voltage V_{OH} is lower than the voltage E_{OH}. If the voltage across the ideal diode is reversed, then it behaves like an open circuit. Similarly for diode D_L. From this model the voltage (in millivolts) can be obtained from the current (in milliamperes) to an accuracy of $\pm$10mV by using the relationships

$$\text{For } V_{OH} : V_O = -\ 850 - 6I_O \qquad 8\,\text{mA} < I_O < 40\,\text{mA} \tag{A.1}$$

$$\text{For } V_{OL} : V_O = -1670 - 8I_O \qquad 2\,\text{mA} < I_O < 16\,\text{mA} \tag{A.2}$$

In figure A.2 we see that the input impedance of ECL consists of a 50 kΩ resistance. The choice of values for the resistance R_T, shown in figure A.3, is covered in chapter 6. It turns out that its value is usually two or more orders of magnitude smaller than the 50 kΩ device input-impedance. Accordingly the 50 kΩ input resistance will be disregarded in the rest of this treatment.

The input terminals of the logic device also have a parasitic capacitance of 2 pF which appears in parallel with the 50 kΩ resistance. This

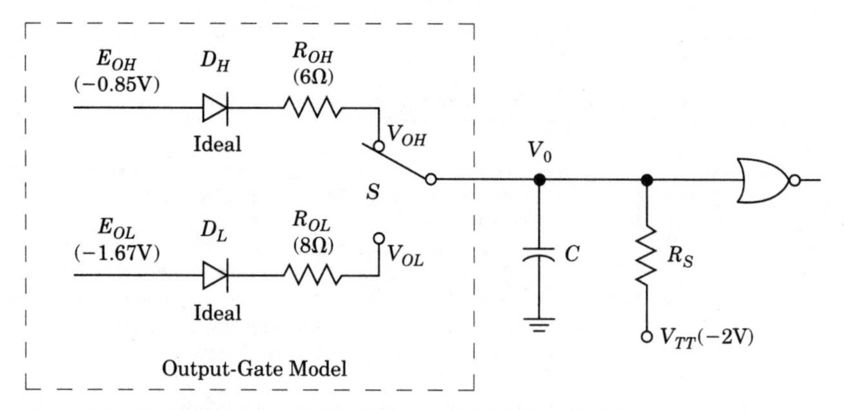

Figure A.3 Model for determining the output behavior of ECL.

is shown in figure A.3 as the capacitor C. This capacitance affects the design of the interconnections between ECL logic devices. This is a topic that is addressed in chapter 3. It will be seen in the material that follows that the presence of this capacitance creates the need for the pulldown arrangement consisting of the voltage V_{TT} and the resistor R_T.

A.2 Requirement for Pulldown Resistors

It will be assumed that the driving gate shown on the left of figure A.3 has gone from a high-output state to a low-output state. This is modeled by the switch S being thrown to the lower position. In the absence of R_T, when the gate output goes from E_{OH} to E_{OL}, the capacitor would hold the higher output voltage V_{OH}, and would never readjust to the new lower voltage V_{OL} because D_L would be back-biased after the gate output-transition. The pulldown resistor R_T is needed as a leakage path to allow the voltage V_O on capacitor C to readjust downward within a reasonable amount of time. The pulldown resistor works in conjunction with the pulldown voltage V_{TT}. This is chosen to lie slightly below E_{OL}. If it is set at too low a value, then the power dissipation in R_T becomes excessive.

To get a feeling of what happens when a transition occurs from the higher output-voltage E_{OH} to the lower output-voltage E_{OL}, we find, using the superposition theorem, that the output voltage prior to the transition is

$$V_O = V_{OH} = \frac{R_T}{R_{OH} + R_T} E_{OH} + \frac{R_{OH}}{R_{OH} + R_T} V_{TT} \tag{A.3}$$

After the input transition, the model is the same as shown in figure A.3, but the switch S is in the lower position. The ideal diode D_L is cut off since the capacitor is still holding the voltage V_{OH} on the diode's right side which exceeds the voltage E_{OL} on the diode's left side. The capacitor C starts at a voltage of V_{OH} and proceeds exponentially toward the voltage V_{TT} with a time constant

$$t_{c1} = R_T C \tag{A.4}$$

The time dependence is given by

$$v_O(t) = [V_{TT} - (V_{TT} - V_{OH})e^{-t/t_{c1}}]u(t) \tag{A.5}$$

The diode D_L remains cut off until the voltage on its right side drops to E_{OL} at time t_1. At this time the diode D_L turns on, and the voltage on the capacitor C starts at E_{OL} and proceeds to the steady-state value V_{OL} which is given by

$$V_O = V_{OL} = \frac{R_T}{R_{OL} + R_T} E_{OL} + \frac{R_{OL}}{R_{OL} + R_T} V_{TT} \tag{A.6}$$

with a time constant

$$t_{c2} = (R_{OL} || R_T)C \approx R_{OL}C \ll t_{c1} \tag{A.7}$$

The last is true because in practice

$$R_{OL} \ll R_T \tag{A.8}$$

Similarly to (A.5), the expression for voltage during this second transition is

$$v_O(t) = \left[V_{OL} - (V_{OL} - E_{OL})e^{-(t-t_1)/t_{c2}} \right] u(t - t_1) \tag{A.9}$$

Example A.1 Assume that $C = 2\,\text{pF}$ and we have arbitrarily chosen a value of $70\,\Omega$ for R_T. For the parameters shown in figure A.3 we get $V_{OH} = -0.941\,\text{V}$. After the gate makes the transition from the high to the low state, we find from (A.3) that $t_{c1} = 140\,\text{ps}$ and from (A.5) we get

$$v_O(t) = \left[-2 + 1.059e^{-t/140} \right] u(t)$$

The output has the above time dependence until $v_O(t)$ reaches the value $E_{OL} = -1.67\,\text{V}$. This occurs at time $t_1 = 163\,\text{ps}$. Diode D_L then turns on and the time constant, as given by (A.7), changes to $t_{c2} = 16\,\text{ps}$. The voltage then follows the time dependence of (A.9), which for our case is

$$v_O(t) = \left[-1.704 + 0.034e^{-(t-163)/16} \right] u(t - 163)$$

This is within 5 mV of the steady-state value after 31 ps, so we can conclude that the whole process is finished in less than 200 ps. ■

Bibliography

1. *F100K ECL Logic Databook and Design Guide,* 1990 Edition, National Semiconductor Corporation, Santa Clara, California.
2. W. R. Blood, Jr., *MECL System Design Handbook,* Fourth Edition, Motorola Semiconductor Products Inc., Phoenix, Arizona, 1988.

Problems

PA.1 Following (A.2) it is pointed out that the $50\,k\Omega$ internal input impedance of ECL gates is much greater than the value of the external resistance R_T, hence the former is disregarded in the analysis. If the $50\,k\Omega$ impedance were the only impedance connected, and it interacted with the parasitic capacitance of 2 pF which appears in parallel with it, what would be the total elapsed time required for the settling of the voltage for the problem in example 5.1? How does this compare with the ECL gate propagation delay of less than 1 ns?

PA.2 In connection with figure A.3, assume that $R_T = 270\,\Omega$, $C = 2\,pF$ and that $V_{TT} = -4.5\,V$.

(a) Write the expression for $v_O(t)$ and find the time needed for the voltage to reach $-1.67\,V$.

(b) Write the time expression for the continuation of the charging process, and determine the total time required to reach within 5 mV of the steady state.

(c) Plot the above results.

PA.3 Assume that the circuit in figure A.3 is in the V_{OL} steady state. The gate output makes a transition to the high logic state. Find the amount of time necessary for the output to reach within 5 mV of the steady-state value V_{OH}.

PA.4 It is apparent from example A.1 that the time required for a gate output to make a transition is small when compared to a toggling rate of 1 ns.

(a) Find an expression for the power dissipated by the resistor R_T in figure A.3 on the assumption that gates change binary states every nanosecond.

(b) Evaluate the expression for $R_T = 25\,\Omega$, $50\,\Omega$, and $80\,\Omega$. V_{TT} is fixed at $-2\,V$.

(c) It will be assumed that a $-2\,V$ power supply for V_{TT} is not readily available but there is a $-4.5\,V$ power supply on hand. It is desired to replace the resistor R_T and the voltage supply V_{TT} by a resistor R_1 connected from the $-4.5\,V$ supply to the gate input, and a resistor R_2 connected from the gate input to ground. Find the values of R_1 and R_2 so that we have a Thevenin equivalent for each case in part (b).

(d) Find the expression for the power dissipated in the resistors in part (c) and compare with the results of part (b). It is assumed that the circuit spends 50%of its time in the high state and 50% of its time in the low state.

Appendix

B

Laplace Transform Review

It is assumed that the reader of this book has had a background in Laplace transforms. The material presented here is meant to give a short and painless review of the concepts necessary for the understanding of the Laplace transform analysis which is needed for this book. The material is presented in a logical way in spite of the fact that some of the rigor is dispensed with. For a more thorough treatment the reader is advised to consult other references [1–3].

B.1 The Laplace Transform

Let $x(t)$ be a time function that is zero for negative values of t, namely

$$x(t) = 0, \qquad \text{for } t < 0 \tag{B.1}$$

The Laplace transform of $x(t)$, which is designated by $\mathscr{L}[x(t)]$, is the function $X(s)$ defined by

$$\mathscr{L}[x(t)] \equiv \int_{0-}^{\infty} x(t)e^{-st}dt = X(s) \tag{B.2}$$

The use of the lower limit of integration in (B.2) requires some explanation [Ref. 1, p. 172]. Usually when dealing with electric circuits, one uses $t = 0$ as the instant of time at which an important event (such as the closing of a switch) takes place. An instant prior to $t = 0$ is designated by $t = 0-$, and an instant following $t = 0$ is designated by $t = 0+$. If there is no discontinuity in the function $x(t)$ at $t = 0$, then the lower limit of integration can be $0-$, 0, or $0+$ without affecting the outcome. But if there is an impulse at $t = 0$, then it is best to use the lower limit appearing in (B.2), so that the impulse will be included. Hence (B.2) represents the most universally useful definition for the Laplace transform when dealing with circuits in which critical events take place at $t = 0$.

Integration is a linear process. It can therefore be proven very readily (as an exercise) that the Laplace transform is linear. Hence, the Laplace transform of the linear combination of two functions is the linear combination of the transforms of the functions. To get a feeling for the use of (B.2) we consider a few useful time functions. We start with the unit step function $u(t)$, defined by

$$u(t) \equiv \begin{cases} 0, & \text{for } t < 0 \\ 1, & \text{for } 0 < t \end{cases} \tag{B.3}$$

shown in figure B.1.

We will obtain the Laplace transform of the step function indirectly by first finding the Laplace transform of the exponentially decreasing function

$$x(t) = e^{-at}u(t) \tag{B.4}$$

Substituting (B.4) into (B.2) we readily obtain

$$\mathscr{L}[e^{-at}u(t)] = \frac{1}{s + a} \tag{B.5}$$

If we allow the parameter a to go to zero in (B.5), then we have the Laplace transform of the unit step function

$$\mathscr{L}[u(t)] = \frac{1}{s} \tag{B.6}$$

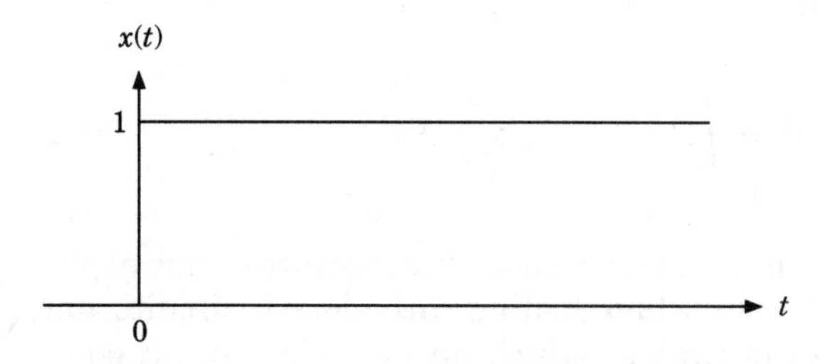

Figure B.1 Unit step function $u(t)$.

Another function of immediate interest is the *unit impulse* $\delta(t)$ which is also referred to as the *delta function*. It has the properties

$$\int_a^b \delta(t)\,dt \equiv \begin{cases} 0, & \text{if interval } (a,b) \text{ excludes } t = 0 \\ 1, & \text{if interval } (a,b) \text{ includes } t = 0 \end{cases} \tag{B.7}$$

The delta function can be conceptualized as a very narrow pulse of infinite height occurring at $t = 0$. Its dimensions have to be very carefully controlled for it to have the area of unity stipulated by (B.7). A time function that produces a delta function in the limit is shown in figure B.2. It can be shown that the time derivative of a unit step function is the unit impulse.

It can be readily verified, by substitution into (B.2), that the unit impulse has a Laplace transform of unity, namely

$$\mathscr{L}[\delta(t)] = 1 \tag{B.8}$$

A few theorems will prove useful for building up a table of some common Laplace transform pairs.

Theorem B.1

If $x(t)$ has the Laplace transform $X(s)$, then the damped time function $e^{-at}x(t)$ has the Laplace transform

$$\mathscr{L}[e^{-at}x(t)] = X(s + a) \tag{B.9}$$

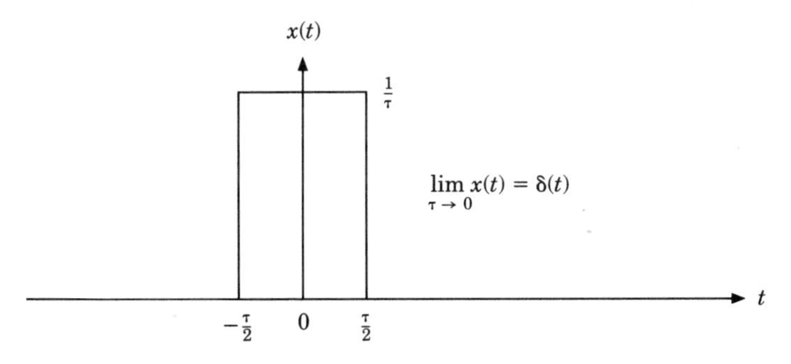

Figure B.2 Example of a function producing a unit impulse.

Proof

Substitute $e^{-at}x(t)$ into (B.2) to obtain

$$\mathcal{L}[e^{-at}x(t)] = \int_{0-}^{\infty} [e^{-at}x(t)]e^{-st}dt = \int_{0-}^{\infty} x(t)e^{-(s+a)t}dt$$

Comparing the last expression on the right with (B.2) we see that it represents $X(s + a)$, hence (B.9) is proven. ■

Example B.1 We will use the above theorem to find the Laplace transforms of sine and cosine functions.

Applying (B.9) to (B.6) we readily find

$$\mathcal{L}[e^{j\omega_0 t}u(t)] = \frac{1}{s - j\omega_0} \tag{B.10}$$

Using the well-known formula due to Euler

$$Ve^{jq} = V(\cos q + j \sin q) \tag{B.11}$$

on the left side of (B.10), and rationalizing the right side of (B.10) we get

$$\mathcal{L}[\cos \omega_0 t\, u(t) + j \sin \omega_0 t\, u(t)] = \frac{s}{s^2 + \omega_0^2} + \frac{j\omega_0}{s^2 + \omega_0^2} \tag{B.12}$$

Equating real and imaginary parts, we obtain the two transforms

$$\mathcal{L}[\cos \omega_0 t\, u(t)] = \frac{s}{s^2 + \omega_0^2} \tag{B.13}$$

$$\mathcal{L}[\sin \omega_0 t\, u(t)] = \frac{\omega_0}{s^2 + \omega_0^2} \tag{B.14}$$

■

Theorem B.2

If $x(t)$ has the Laplace transform $X(s)$, then

$$\mathcal{L}[(-t)^n x(t)] = \frac{d^n}{ds^n}X(s) \tag{B.15}$$

Proof: Start with (B.2) as below

$$X(s) = \int_{0-}^{\infty} x(t)e^{-st}dt = \mathscr{L}[x(t)]$$

and successively differentiate the above expression with respect to s. Each differentiation under the integral brings down a term $-t$ from the exponential. After n differentiations we have

$$\frac{d^n}{ds^n}X(s) = \int_{0-}^{\infty} (-t)^n x(t)e^{-st}dt = \mathscr{L}[(-t)^n x(t)]$$

which proves (B.15). ■

Example B.2 We will use the above theorem to find the Laplace transform of $t^n u(t)$.

Applying (B.15) to (B.6) we readily find that

$$\mathscr{L}[t^n u(t)] = \frac{n!}{s^{n+1}} \tag{B.16}$$

■

Theorem B.3

The time shifting theorem states that if the Laplace transform of $x(t)u(t)$ is $X(s)$, then

$$\mathscr{L}[x(t - t_0)u(t - t_0)] = e^{-t_0 s}X(s) \tag{B.17}$$

Proof

Substitute the function $x(t - t_0)u(t - t_0)$ into (B.2). This function does not exist prior to $t = t_0-$, hence the latter becomes the lower limit of integration.

$$\mathscr{L}[x(t - t_0)u(t - t_0)] = \int_{t_0-}^{\infty} x(t - t_0)e^{-st}dt$$

Change variables in the above by substituting ξ for $t - t_0$ to obtain

$$\mathscr{L}[x(t - t_0)u(t - t_0)] = \int_{0-}^{\infty} x(\xi)e^{-s(\xi+t_0)}d\xi$$

The exponential involving t_0 is not involved in the integration, so it can be brought outside the integral, with the final step of the proof

$$\mathscr{L}[x(t - t_0)u(t - t_0)] = e^{-t_0 s}\int_{0-}^{\infty} x(\xi)e^{-s\xi}d\xi$$

which proves (B.17). ■

Example B.3 We will use the above theorem to find the Laplace transform of the function shown in figure B.3.

The function $x(t)$ shown in figure B.3 can be considered the superposition of step functions as

$$x(t) = u(t) - \frac{1}{2}u(t-1) - \frac{1}{2}u(t-2) \tag{B.18}$$

Applying (B.17) to (B.18), we write without any difficulty

$$X(s) = \frac{1}{s} - \frac{1}{2s}e^{-s} - \frac{1}{2s}e^{-2s} = \left[1 - \frac{1}{2}e^{-s} - \frac{1}{2}e^{-2s}\right]\frac{1}{s} \tag{B.19}$$

Although it was not explicitly mentioned above, the linearity property was used in going from (B.18) to (B.19). ■

Theorem B.4

The differentiation theorem states that if the Laplace transform of $x(t)u(t)$ is $X(s)$, then

$$\mathcal{L}\left[\frac{dx(t)}{dt}\right] = sX(s) - x(0-) \tag{B.20}$$

where, as was stated previously, $x(0-)$ represents the value of $x(t)$ an instant before $t = 0$.

Figure B.3 A function of time.

Proof

Start with (B.2)

$$\mathcal{L}[x(t)] = \int_{0-}^{\infty} x(t)e^{-st}dt = X(s)$$

Integrate the above equation by parts assigning

$$u = x(t) \qquad du = \left[\frac{dx(t)}{dt}\right]dt$$

$$dv = e^{-st}dt \qquad v = -\frac{1}{s}e^{-st}$$

to obtain

$$X(s) = -\frac{1}{s}x(t)e^{-st}\Big|_{0-}^{\infty} + \frac{1}{s}\int_{0-}^{\infty}\left[\frac{dx(t)}{dt}\right]e^{-st}dt$$

Evaluating the above at the limits and closely examining the integral, we find that the above is equivalent to

$$X(s) = \frac{x(0-)}{s} + \frac{1}{s}\mathcal{L}\left[\frac{dx(t)}{dt}\right]$$

Rearranging the above expression we readily get (B.20). ■

Example B.4 From (B.16) we know that

$$\mathcal{L}[tu(t)] = \frac{1}{s^2}$$

and also that the step function is the derivative of the ramp

$$u(t) = \frac{d}{dt}[tu(t)]$$

We can therefore apply (B.20) to find that

$$\mathcal{L}[u(t)] = s\left[\frac{1}{s^2}\right] - tu(t)\Big|_{0-} = \frac{1}{s}$$

which is just another method of computing the result of (B.6). ■

Theorem B.5

We will use the notation

$$x^{-1}(t) = \int_{-\infty}^{t} x(\tau)d\tau \tag{B.21}$$

The integration theorem states that if the Laplace transform of $x(t)u(t)$ is $X(s)$, then

$$\mathscr{L}\left[\int_{-\infty}^{t} x(\tau)d\tau\right] = \frac{X(s) + x^{-1}(0-)}{s} \tag{B.22}$$

where from (B.21) it is understood that

$$x^{-1}(0-) = \int_{-\infty}^{0-} x(\tau)d\tau \tag{B.23}$$

The above is an initial condition representing the contribution of $x(t)$ to the integral for the time preceding $t = 0$. The above term is zero for functions satisfying (B.1).

Proof

Start with (B.20) in the form

$$\mathscr{L}\left[\frac{dy(t)}{dt}\right] = sY(s) - y(0-)$$

and substitute

$$y(t) = x^{-1}(t)$$

to obtain

$$\mathscr{L}[x(t)] = s\mathscr{L}[x^{-1}(t)] - x^{-1}(0-)$$

Rearranging the above expression we obtain

$$\mathscr{L}[x^{-1}(t)] = \frac{\mathscr{L}[x(t)] + x^{-1}(0-)}{s}$$

which is just (B.22) with different notation. ■

TABLE B.1 A Short Table of Laplace Transforms.

Name	Time function	Laplace transform
Unit impulse	$\delta(t)$	1
Unit step	$u(t)$	$\frac{1}{s}$
Unit ramp	$tu(t)$	$\frac{1}{s^2}$
n*th* order ramp	$t^n u(t)$	$\frac{n!}{s^{n+1}}$
Exponential	$e^{-at}u(t)$	$\frac{1}{s+a}$
Damped unit ramp	$te^{-at}u(t)$	$\frac{1}{(s+a)^2}$
Damped n*th* order ramp	$t^n e^{-at}u(t)$	$\frac{n!}{(s+a)^{n+1}}$
Sine wave	$\sin \omega_0 t u(t)$	$\frac{\omega_0}{s^2+\omega_0^2}$
Cosine wave	$\cos \omega_0 t u(t)$	$\frac{s}{s^2+\omega_0^2}$
Damped sine wave	$e^{-at}\sin \omega_0 t u(t)$	$\frac{\omega_0}{(s+a)^2+\omega_0^2}$
Damped cosine wave	$e^{-at}\cos \omega_0 t u(t)$	$\frac{s+a}{(s+a)^2+\omega_0^2}$

Example B.5 The last theorem will be used to find the Laplace transform of the *integral* of $e^{-at}u(t)$. Starting with (B.5) we have

$$\mathscr{L}[e^{-at}u(t)] = \frac{1}{s+a}$$

The function $e^{-at}u(t)$ is zero for negative time. As a consequence $x^{-1}(0-)$ will be zero. Hence substitution of the last equation into (B.22) produces the result

$$\mathscr{L}\left[\int_{-\infty}^{t} e^{-a\tau}u(\tau)d\tau\right] = \frac{1}{s(s+a)}$$ ■

A short table of Laplace transforms is given in table B.1. This is followed by a short table of Laplace transform theorems appearing in table B.2.

B.2 Inversion of Laplace Transforms

Consider a Laplace transform that is the ratio of two polynomials as

$$X(s) = \frac{N(s)}{D(s)} = \frac{a_m s^m + a_{m-1}s^{m-1} + \cdots + a_1 s + a_0}{s^n + b_{n-1}s^{n-1} + \cdots + b_1 s + b_0} \tag{B.24}$$

TABLE B.2 A Short Table of Laplace Transform Theorems.

Definitions

$$\mathscr{L}[x(t)] = \int_{0-}^{\infty} x(t)e^{-st}\,dt = X(s)$$

$$x^{-1}(t) = \int_{-\infty}^{t} x(\tau)\,d\tau$$

Name	Theorem
Linearity	$\mathscr{L}[ax_1(t) + bx_2(t)] = aX_1(s) + bX_2(s)$
Damping	$\mathscr{L}[e^{-at}x(t)] = X(s + a)$
Multiplication by polynomial	$\mathscr{L}[(-t)^n x(t)] = \frac{d^n}{ds^n} X(s)$
Time shifting	$\mathscr{L}[x(t - to)u(t - to)] = e^{-tos}X(s)$
Time differentiation	$\mathscr{L}\left[\frac{dx(t)}{dt}\right] = sX(s) - x(0-)$
Time integration	$\mathscr{L}[x^{-1}(t)] = \frac{X(s) + x^{-1}(0-)}{s}$

We wish to determine the time function $x(t)$, whose Laplace transform is $X(s)$.

Functions of *s* with Simple Poles The simplest case to consider is that of functions of s with poles of first order only. Assuming that the poles are located at $p_1, p_2, \ldots, p_n$, then $X(s)$ can be written in the form

$$X(s) = \frac{N(s)}{(s - p_1)(s - p_2)\cdots(s - p_n)} \tag{B.25}$$

To find the time function $x(t)$, $X(s)$ is expanded into a partial-fraction expansion of the form

$$X(s) = \frac{A_1}{s - p_1} + \frac{A_2}{s - p_2} + \cdots + \frac{A_n}{s - p_n} \tag{B.26}$$

where the coefficients A_i are found using the equation

$$A_i = (s - p_i)X(s)|_{s=p_i} \tag{B.27}$$

We restate (B.5), namely that the Laplace transforms of exponentials give rise to poles

$$\mathcal{L}[e^{-pt}u(t)] = \frac{1}{s+p} \tag{B.28}$$

From the above it is apparent that the inverse Laplace transform of (B.26) is given by

$$x(t) = A_1 e^{p_1 t} + A_2 e^{p_2 t} + \cdots + A_n e^{p_n t} \tag{B.29}$$

Example B.6 It is desired to find the time function $x(t)$ that corresponds to the Laplace transform

$$X(s) = \frac{9(s^2 + 7s + 10)}{(s+1)(s+4)(s+7)} \tag{B.30}$$

As a first step it is desired to put (B.30) into the form shown in (B.26). By the repeated application of (B.27) we find that $A_1 = 2$, $A_2 = 2$, and $A_3 = 5$. The function $X(s)$ can therefore be written as an equivalent partial-fraction expansion

$$X(s) = \frac{2}{s+1} + \frac{2}{s+4} + \frac{5}{s+7} \tag{B.31}$$

Using (B.28) it is readily apparent that the function $x(t)$, which is the inverse transform of $X(s)$, is given by

$$x(t) = [2e^{-t} + 2e^{-4t} + 5e^{-7t}]u(t) \tag{B.32}$$

■

Functions of *s* with Multiple Poles If the Laplace transform $X(s)$ has a pole of multiplicity r at p_j as shown below

$$X(s) = \frac{N(s)}{(s-p_1)(s-p_2)\cdots(s-p_j)^r} \tag{B.33}$$

then it can be written in a partial-fraction expansion of the form

$$X(s) = \frac{A_1}{s-p_1} + \frac{A_2}{s-p_2} + \cdots + \frac{A_{j1}}{s-p_j} + \frac{A_{j2}}{(s-p_j)^2} + \cdots + \frac{A_{jr}}{(s-p_j)^r} \tag{B.34}$$

The numerator coefficients for the simple poles can be evaluated using (B.27), but for the multiple poles we use

$$A_{jn} = \frac{1}{(r-n)!}\frac{d^{r-n}}{ds^{r-n}}(s-p_j)^r X(s)|_{s=p_j} \tag{B.35}$$

The inverse transform for the simple pole terms is found using (B.28). For the inverse transform of the higher order pole terms we use

$$\mathcal{L}\left[\frac{t^{n-1}}{(n-1)!}e^{pt}u(t)\right] = \frac{1}{(s-p)^n} \tag{B.36}$$

It is readily apparent from the above that multiple poles produce polynomials multiplied by exponentials.

Example B.7 The application of the above is very simple. Take as an example

$$X(s) = \frac{s^2+1}{s(s+1)^3} = \frac{A}{s} + \frac{A_{11}}{s+1} + \frac{A_{12}}{(s+1)^2} + \frac{A_{13}}{(s+1)^3} \tag{B.37}$$

The simple pole technique of (B.27) is used to find that $A = 1$. To find the other numerator coefficients of the partial-fraction expansion, we form the expression

$$(s+1)^3 X(s) = s + \frac{1}{s} = \frac{A}{s}(s+1)^3 + A_{11}(s+1)^2 + A_{12}(s+1) + A_{13} \tag{B.38}$$

The value of A_{13} is obtained from (B.38) by setting $s = -1$. It is readily apparent that $A_{13} = -2$. Next (B.38) is differentiated with respect to s, to obtain

$$\frac{d}{ds}(s+1)^3 X(s) = 1 - \frac{1}{s^2} = A\frac{2s-1}{s^2}(s+1)^2 + 2A_{11}(s+1) + A_{12} \tag{B.39}$$

When we set $s = -1$, we obtain $A_{12} = 0$. Differentiating (B.38) again with respect to s, we get

$$\frac{d^2}{ds^2}(s+1)^3 X(s) = \frac{2}{s^3} = 2A\frac{s^2-s+1}{s^3}(s+1) + 2A_{11} \tag{B.40}$$

Setting $s = -1$ it follows that $A_{11} = -1$. The partial-fraction expansion for (B.37) therefore takes the form

$$X(s) = \frac{1}{s} - \frac{1}{s+1} - \frac{2}{(s+1)^3} \tag{B.41}$$

The above is inverted using (B.36), and the result is

$$x(t) = [1 - (1+t^2)e^{-t}]u(t) \tag{B.42}$$

■

Functions of s with Complex Conjugate Poles The partial-fraction expansion treatment when complex conjugate poles occur in $X(s)$ is the same as it is for real poles, but the resulting time function is somewhat different. Since the polynomial coefficients in $X(s)$ are real, then complex poles must occur in conjugate pairs. If p_i is a complex pole, then a portion of $X(s)$ will be the partial response $X_i(s)$ which will have the form

$$X_i(s) = \frac{A_i}{s - p_i} + \frac{A_i^*}{s - p_i^*} \tag{B.43}$$

The asterisk in the above equation signifies the complex conjugate. The right side of the last equation produces a real $X_i(s)$ because of the simple fact that the sum of a complex number and its conjugate equals twice the real part of either the number or the conjugate, as stated by

$$A + A^* = 2\text{Re}\,(A) = 2\text{Re}\,(A^*) \tag{B.44}$$

where the notation Re($\cdot$) in the last equation indicates the real part of the argument.

The inverse Laplace transform of (B.43) is obtained using (B.28), with the result

$$x_i(t) = A_i e^{p_i t} + A_i^* e^{p_i^* t} \tag{B.45}$$

Applying (B.44) to the above we have

$$x_i(t) = 2\text{Re}\left[A_i e^{p_i t}\right] \tag{B.46}$$

Writing the complex A_i in terms of magnitude and phase

$$A_i = |A_i| e^{j\theta_i} \tag{B.47}$$

and the complex p_i in terms of its real and imaginary parts

$$p_i = \sigma_i + j\omega_i \tag{B.48}$$

allows (B.46) to be rewritten as

$$x_i(t) = 2\text{Re}\left[\,|A_i| e^{j\theta_i} e^{(\sigma_i + j\omega_i)t}\right] \tag{B.49}$$

or

$$x_i(t) = 2\text{Re}\left[\,|A_i| e^{\sigma_i t} e^{j(\omega_i t + \theta_i)}\right] \tag{B.50}$$

To put the above into real form we make use of Euler's formula of (B.11) to get the final result

$$x_i(t) = 2|A_i| e^{\sigma_i t} \cos(\omega_i t + \theta_i) \tag{B.51}$$

Example B.8 As an example, take

$$X(s) = \frac{3s^2 + 22s + 29}{(s + 1)(s + 2 + j3)(s + 2 - j3)} \tag{B.52}$$

Using (B.27) to find the partial-fraction expansion, we obtain

$$X(s) = \frac{1}{s + 1} + \frac{1 + j2}{s + 2 + j3} + \frac{1 - j2}{s + 2 - j3} \tag{B.53}$$

In finding the numerator coefficients of this partial-fraction expansion, only two had to be found, the last being the conjugate of the previous. Using (B.34), the time function $x(t)$ is expressed as

$$x(t) = \left[e^{-t} + (1 + j2)e^{-(2+j3)t} + (1 - j2)e^{-(2-j3)t}\right]u(t) \tag{B.54}$$

According to (B.44) the last can be written in the form

$$x(t) = \left\{e^{-t} + 2\text{Re}\left[2.24e^{j1.11}e^{-(2+j3)t}\right]\right\}u(t) \tag{B.55}$$

where $2.24e^{j1.11}$ is merely $1 + j2$ written in polar form. Now Euler's formula of (B.11) is used to obtain the final real form

$$x(t) = [e^{-t} + 4.48e^{-2t}\cos(3t - 63.4°)]u(t) \tag{B.56}$$

■

B.3 Application of Laplace Transforms to Circuits

Now that we know how to find Laplace transforms from time functions and how to invert Laplace transforms to find the corresponding time functions, it remains to demonstrate how to apply them to circuits.

On the left side of figure B.4 we have a capacitor shown with an initial voltage of value V_0. In the first line of the center of the diagram we have the relationship between current and voltage for a capacitor.

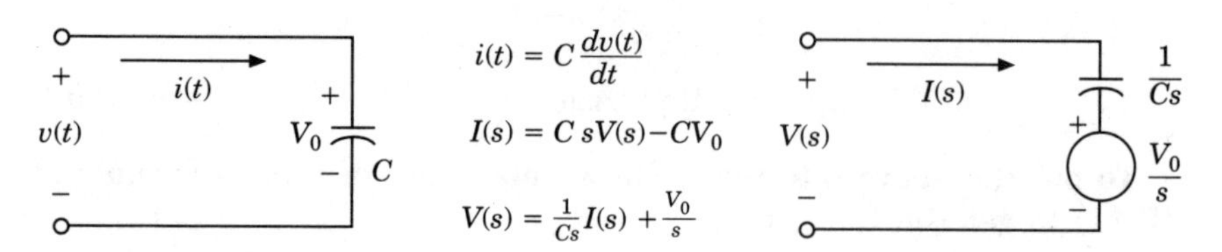

Figure B.4 Laplace transform representation for capacitors.

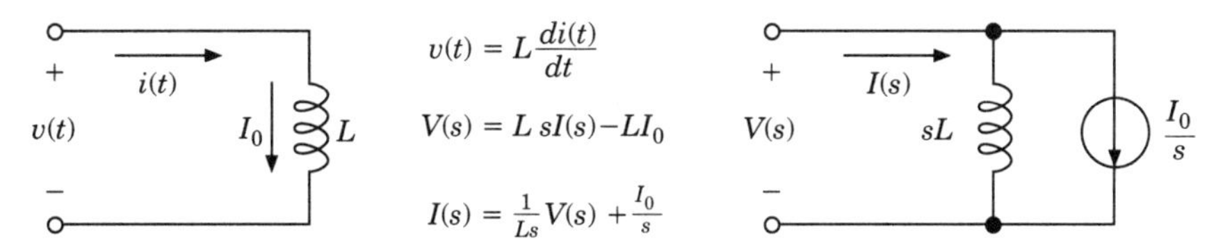

Figure B.5 Laplace transform representation for inductors.

The second line shows the transform of the first line. The third line is merely a more useful restatement of the second line, from which we obtain the Laplace transform representation of a capacitor shown in the diagram on the right.

On the left side of figure B.5 we have an inductor shown with an initial current of value I_0. In the first line of the center of the diagram we have the relationship between voltage and current for an inductor. The second line shows the transform of the first line. The third line is merely an algebraic restatement of the second line, from which we obtain the Laplace transform representation of an inductor shown in the diagram on the right.

Example B.9 As an example, we will find the output voltage $v_a(t)$ for the circuit shown in figure B.6a. The capacitor has no initial voltage prior to the time the switch is thrown.

We proceed to a solution of the problem by replacing the entire circuit on the left with its Laplace transform equivalent as shown in figure B.6b. The expression for $V_a(s)$, is found directly from the circuit in figure B.6b.

$$V_a(s) = \frac{1/sC}{R + 1/sC}\left(\frac{V}{s}\right)$$

The above is then written as the ratio of two polynomials in the more convenient form

$$V_a(s) = \frac{(1/RC)\,V}{s(s + 1/RC)}$$

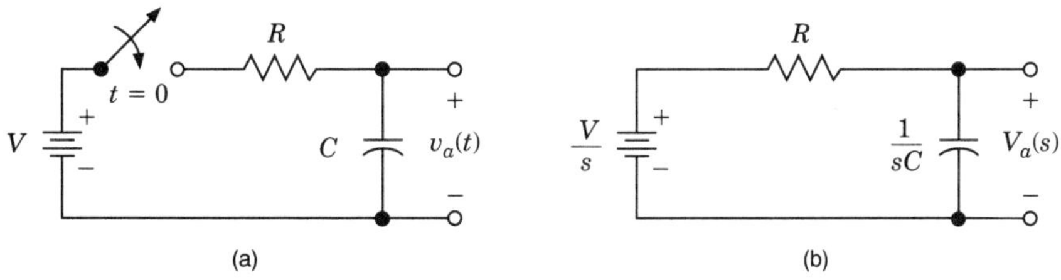

(a) (b)

Figure B.6 An R–C circuit transient problem.

The last expression is expanded into a partial-fraction expansion

$$V_a(s) = \frac{V}{s} - \frac{V}{s + 1/RC}$$

from which it follows that

$$v_a(t) = V\left[1 - e^{-t/RC}\right]u(t)$$ ■

B.4 Conclusion

This concludes the short review of Laplace transforms. The readers should at this point have at their disposal a sufficient background in the subject to readily follow the transient problem material presented in this book. If a deeper understanding of Laplace transforms is desired, then, as was mentioned earlier, other materials [1–3] can be consulted.

References

1. M. E. Van Valkenburg, *Network Analysis*, Third Edition, Prentice Hall, Inc., Englewood Cliffs, New Jersey, 1974.
2. D. K. Cheng, *Analysis of Linear Systems*, Addison-Wesley, Reading, Massachusetts, 1959.
3. R. V. Churchill and J. W. Brown, *Complex Variables and Applications*, McGraw-Hill, New York, 1976.

Problems

PB.1 The Laplace transform has the linearity property because it is based on an integral. Prove by going back to (B.2) that if $\mathcal{L}[x_1(t)] = X_1(s)$ and $\mathcal{L}[x_2(t)] = X_2(s)$, then

$$\mathcal{L}[ax_1(t) + bx_2(t)] = aX_1(s) + bX_2(s)$$

PB.2 Apply superposition to find the Laplace transform of the waveform shown in figure B.7.

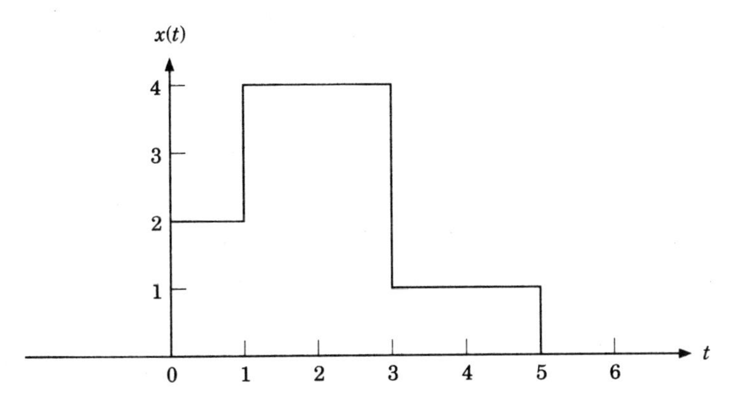

Figure B.7 Composite time function.

PB.3 Verify (B.5) by direct integration.

PB.4 Figure B.8 shows a step function with a ramp transition. Draw the derivative of this function and show that the area under the derivative is unity. Does this hold irrespective of τ?

PB.5 It is well known that the derivative of a step function is a delta function. Differentiate the wave shown in figure B.8. For the derivative

(a) Use the time shifting theorem and (B.3) to find an expression for its Laplace transform.
(b) Let $\tau \to 0$ to find the Laplace transform of $\delta(t)$ indirectly.
(c) Find the Laplace transform of $\delta(t)$ directly from its definition given in (B.7).

PB.6 Use (B.16) to list the Laplace transforms for $t^n u(t)$ for a few values of n.

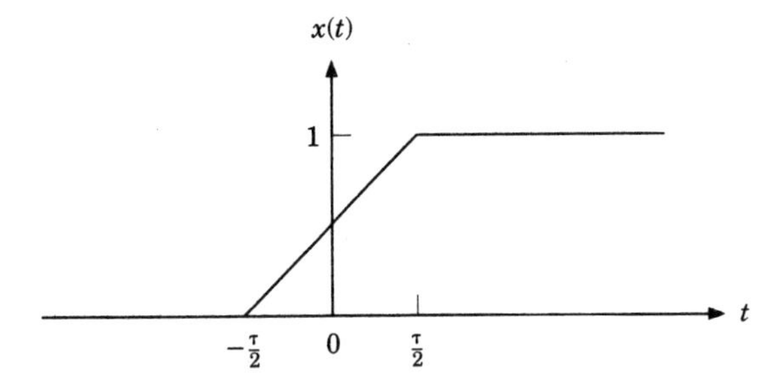

Figure B.8 Step function with ramp transition.

PB.7 Combine (B.9) with (B.16) to find the expression for the Laplace transform of $t^n e^{-at} u(t)$.

PB.8 By writing A in complex rectangular form, prove (B.44).

PB.9 Find the inverse Laplace transforms for

(a) $$X(s) = \frac{(s+4)(s+7)}{(s+1)(s+2)(s+3)(s+5)}$$

(b) $$X(s) = \frac{1}{(s+1)^3(s+3)}$$

PB.10 For each of the functions given, find the inverse Laplace transform and write the result in real form

(a) $$X(s) = \frac{(s+4)(s+7)}{(s+1)(s^2+2s+2)}$$

(b) $$X(s) = \frac{s^2}{(s^2+1)^2}$$

(c) $$X(s) = \frac{1}{s^2-2s+5}$$

PB.11 In (B.38) we found the term $(A/s)(s+1)^3$. Setting $s = -1$ made this term and two successive derivatives vanish. Take a function $F(s)$ consisting of a product of two functions of the form

$$F(s) = Q_1(s)(s-p)^n$$

Show that $F(s)$ and $n - 1$ of its derivatives will vanish when we set $s = p$.

PB.12 For the same problem as in example B.9, find the output voltage $v_a(t)$. The capacitor has an initial voltage of V_0 (with the top terminal positive) prior to the time the switch is thrown.

ANSWER: $v_a(t) = [V + (V_0 - V)e^{-t/RC}]u(t)$

PB.13 In the circuit shown in figure B.9 the capacitor has an initial voltage V_0 prior to the time the switch is thrown. Draw the Laplace transform equivalent circuit and find the expression for $I_a(s)$ and $V_a(s)$. Is the denominator polynomial the same in both results?

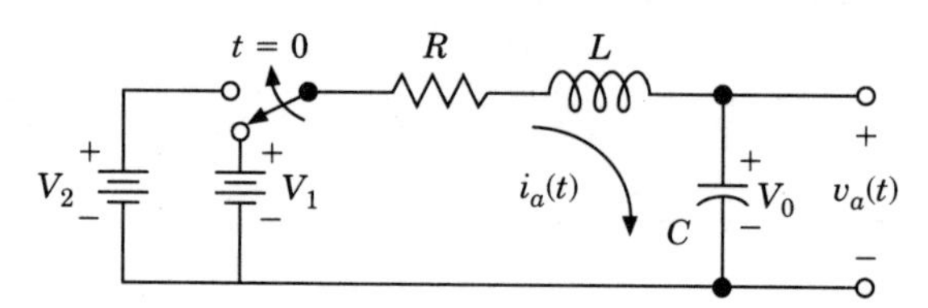

Figure B.9

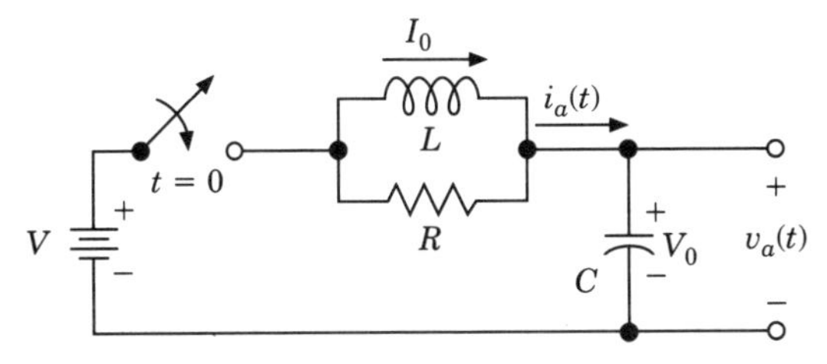

Figure B.10

PB.14 In the circuit shown in figure B.9 the capacitor has an initial voltage V_0 prior to the time the switch is thrown and the inductor has an initial current $i_a(0) = I_0$. Draw the Laplace transform equivalent circuit and find the expression for $I_a(s)$ and $V_a(s)$. Is the denominator polynomial the same in both results?

PB.15 In the circuit shown in figure B.10 the circuit is deenergized prior to the time the switch is thrown. Draw the Laplace transform equivalent circuit and find the expression for $V_a(s)$ and $I_a(s)$. Is the denominator polynomial the same in both results?

PB.16 In the circuit shown in figure B.10 the capacitor has an initial voltage V_0 and the inductor has an initial current I_0 prior to the time the switch is thrown. Draw the Laplace transform equivalent circuit and find the expression for $V_a(s)$.

Index

ABOUT THE AUTHOR

Sol Rosenstark is Professor of Electrical and Computer Engineering at the New Jersey Institute of Technology (NJIT). He previously held positions with Norden Laboratories and AT&T Bell Labs. Dr. Rosenstark has worked in communications theory with special emphasis on spread spectrum communication systems and on command control and communication systems analysis. He has also had a continuing interest in clarifying the theory of feedback amplifiers, and is the author of *Feedback Amplifier Principles*. Dr. Rosenstark has been instrumental in creating the computer engineering program at NJIT, and has written this book for that program.